The Green State

The Green State

Rethinking Democracy and Sovereignty

Robyn Eckersley

The MIT Press
Cambridge Massachusetts
London, England

This book was set in Sabon by SNP Best-set Typesetter Ltd., Hong Kong.
Printed and bound in the United States of America.

Library of Congress Cataloging-in-Publication Data

Eckersley, Robyn, 1958–
The green state : rethinking democracy and sovereignty / Robyn Eckersley.
p. cm.
Includes bibliographical references and index.
ISBN 0-262-05074-9 (alk. paper)—ISBN 0-262-55056-3 (pbk. : alk. paper)
1. Green movement. 2. Democracy. I. Title.
JA75.8.E263 2004
320.5′8–dc22

2003061436

Printed on Recycled Paper.

10 9 8 7 6 5 4 3 2 1

For Eva, for ever

Contents

Preface

This book reflects my attempt to reach beyond the horizons of existing environmental governance, using the current institutions of governance as the point of departure. I have long been inspired by critical theory, and in this project I have sought to enlist, and provide a distinctly green inflection to, critical theory's method of immanent critique.

In placing the state at the center of the analysis, my argument is in some respects unashamedly revisionist given the current shift of academic political focus toward governance without government and the antistatist posture of many radical environmentalists. However, as green parties come in from the periphery and tilt toward the center of political power (e.g., there are encouraging signs in New Zealand and Australia that greens are on the rise), it seems timely to ask how the state might be rescued or perhaps reinvented as a site of democratic public power. Despite the huge transformations wrought by globalization, states still remain gatekeepers of the global order, which seems to me all the more reason to develop a fresh, practical vision of the "good state." In this book I explore what it might take to produce a distinctly *green* democratic state as an alternative to the classical liberal state, the indiscriminate growth dependent welfare state, and the increasingly ascendant neoliberal competition state. This task also entails asking what kind of state or states might facilitate both more active and effective ecological citizenship and more enlightened environmental governance, both domestically and globally. At a minimum a good state would uphold the rule of law and the separation of powers, be free of corruption, and uphold those civil and political rights that are essential to the practice of ecological citizenship. But what else should a *green* state be? What other purposes and roles should it embody and perform? Those few political scientists who addressed this question in the aftermath of the limits-to-growth debate in the early 1970s came up with an eco-authoritarian state. Yet the idea of the state presiding over strict resource and energy rationing, and wide-ranging strictures on consumption, production, population, and technology, seemed anathema to everyone, including many environmentalists. What, then, might a green democratic alternative look like and to what extent, if any, would it differ from the liberal democratic state in terms of its role, rationale, and functions? And what are the prospects of green democratic states emerging in the current, rather inhospitable, global context? In tackling these and related questions, I have drawn on a wide range of disciplines and subdisciplines in the humanities and social sciences, ranging from political theory and sociology to international relations and global political economy, including their budding green offshoots.

This book was written over the past seven years during a transition in my own research focus from the cozy and secluded fold of green political theory to the much more sprawling and complex field of global politics. In making this transition, I am indebted to many colleagues and friends who, by their shining example rather than deliberate effort on their part, drew me into a range of challenging and stimulating intellectual debates that bear upon the future of environmental governance. I wish to single out, in particular, Chris Reus-Smit and Paul James. Chris I heartily thank for introducing me to the constructivist dimension of critical theory, and for his enthusiasm and wholehearted support in my academic journey. Paul I likewise thank for prompting me to think about the "nation" side of the "nation-state" equation, and for his general encouragement in my writing projects. And thanks to both Chris and Paul for serving as critical sounding boards and readers during the production of this manuscript.

Stephanie Trigg, Paul James, and Joel Trigg deserve very special mention as part of our extended family, sharing many meals and much childcare. John Dryzek always lent an ear, provided critical feedback when solicited, was an excellent climbing partner and even better beer brewer. And I thank Rob Watts for managing such incisive comments on top of his very hectic schedule.

I am grateful to the Reshaping Australian Institutions Project at the Australian National University, which enabled me to enjoy a period of research leave in 1996 as a visiting scholar, during which I lay most of the groundwork for this project. I am also indebted to Monash University for three successive Australian Research Council small grants in 1998 through 2000, which gave me the opportunity to continue research on this project. I thank my former colleagues and postgraduates in the Politics Department at Monash University for their friendship and wit over the past decade. I am particularly grateful to Gerry Nagtzaam for his very helpful research assistance over the years and for being such an excellent book scout and Nicole Boldt for library work and help with the bibliography. My academic and administrative colleagues in the Department of Political Science at Melbourne University have provided a friendly and supportive welcome.

It has been a pleasure to work with Clay Morgan at The MIT Press, just as it was when he saw me through *Environmentalism and Political Theory* when he worked for SUNY Press. I am especially grateful to the three anonymous reviewers who offered constructive and perceptive feedback on the manuscript.

Finally, and closer to home, I thank Peter Christoff, my central and most critical of critics on scholarly matters, my nearest and dearest on all other matters, and someone who understands even better than I the trials of completing large projects.

The Green State

1

Introduction

1.1 Why the Green State?

At first glance the notion of a green state might strike many people as a rather quixotic idea, perhaps even a dangerous one. Does it mean a benevolent state, presiding over an ecotopia, the stuff of green dreams? Or does it raise the specter of an authoritarian state, presiding over a strict regime of ecological controls and resource rationing, the stuff of nightmares for liberals? These opposing visions highlight very real divisions among environmentalists, green political theorists, and green party followers about the proper role and future potential of the nation-state in managing ecological problems.[1] Despite the widening ecological critique of the liberal democratic state, the contours of a more constructive green juridical-ethical theory of the state, both domestically and in the context of the state-system and the global order, are not easy to discern. The environmental demands as to what the state ought to be doing (or not doing) in public policy presuppose a more fundamental normative theory of the proper character and role of the nation-state vis-à-vis its own society and territory, the society of states, global civil society, and the global environment. Such a normative theory of the state would need to provide an account of the basis of state legitimacy by developing the regulative ideals that confer authority on, and provide the basis of acceptance of, decisions made in the name of the state. In the past, legitimacy was acquired by the provision of military and domestic security and the regulation and enforcement of contracts. Nowadays that legitimacy is primarily acquired by appeal to democracy, typically representative democracy of the liberal democratic variety. Indeed, the regulative ideals and procedures of liberal democracy provide the most influential yardstick against which alternative normative accounts of the state are usually compared and evaluated. Yet most green political theorists question whether the liberal democratic state is up to the task of steering the economy and society along a genuinely ecologically sustainable path.

This book seeks to develop a political theory of the green state through a series of critical encounters with existing debates about the changing role of the liberal democratic state in an increasingly globalizing world. By "green state" I do not simply mean a liberal democratic state that is managed by a green party government with a set of programmatic environmental goals, although one might anticipate that such a state is most likely to evolve from liberal or social democratic states. Rather, I mean a democratic state whose regulatory ideals and democratic procedures are informed by *ecological* democracy rather than *liberal* democracy. Such a state may be understood as a *postliberal* state insofar as it emerges from an immanent (ecological) critique, rather than from an outright rejection, of liberal democracy.

It was the bourgeoisie who in the eighteenth and nineteenth centuries served as the vanguard for the creation of the liberal democratic state while the labor movement was in the forefront of the social forces that created the social democratic state (or welfare state) in the twentieth century. If a more democratic and outward-looking state—the green democratic state—is ever to emerge in the new millennium, then the environment movement and the broader green movement will most likely be its harbingers. This is unlikely to occur without a protracted struggle. In view of the intensification of economic globalization and the ascendancy of neoliberal economic policy, the challenges are considerable.

This inquiry seeks to confront these challenges and to develop a normative theory of the transnational, green democratic state out of this critical encounter. In developing and defending new regulatory ideals of the green democratic state, and the practice of what might be called "ecologically responsible statehood," this book seeks to connect the moral and practical concerns of the green movement with contemporary debates about the state, democracy, law, justice, and difference. In particular, I seek to outline the constitutional structures of a green democratic state that might be more amenable to protecting nature than the liberal democratic state while maintaining legitimacy in the face of cultural diversity and increasing transboundary and sometimes global ecological problems. I hope to show how a rethinking of the principles of ecological democracy might ultimately serve to cast the state in a new role: that of an ecological steward and facilitator of transboundary democracy rather than a selfish actor jealously protecting its territory and ignoring or discounting the needs of foreign lands. Such a normative ideal poses a fundamental challenge to traditional notions of the nation, of national sovereignty, and the organization of democracy in terms of an enclosed territorial space and polity. It requires new democratic procedures, new decision rules, new forms of political representation and participation, and a more fluid set of relationships and understandings among states and peoples.

My project, then, is clearly to re-invent states rather than to reject or circumvent them. In this respect my inquiry swims against the strong current of scepticism by pluralists, pragmatists, and realists toward "attempts to invest the state with normative qualities, or higher responsibilities to safeguard the public interest, or articulate and uphold a framework of moral rules, or a distinctive sphere of justice."[2] Although historical and critical sociological inquiries into state formation and state practices continue apace, it has become increasingly unfashionable to defend normative theories of the state. Yet these two different approaches cannot be wholly dissociated. As Andrew Vincent reminds us, historical and sociological description and explanation are unavoidably saturated with normative preconceptions, even if they are not always made explicit.[3] And if the traditional repertoire of normative preconceptions about the purposes of the state and the state system is inadequate when it comes to representing ecological interests and concerns, then I believe it has become necessary to invent a new one.

However, any attempt to develop a green theory about the proper role and purpose of the state in relation to domestic and global societies and their environments must take, as its starting point, the current structures of state governance, and the ways in which such structures are implicated in either producing and/or ameliorating ecological problems. This recognition of the important linkages between historical/sociological explanation and normative theory has been one of the hallmarks of

Marxist-inspired critical social theory. Accordingly it has sought to avoid the inherent conservatism of purely positivistic sociological explanation, on the one hand, while avoiding merely wishful utopian dreaming, on the other.[4] Throughout this inquiry, I build on both the method and normative orientation of critical theory. Specifically, I look for emancipatory opportunities that are immanent in contemporary processes and developments and suggest how they might be goaded and sharpened in ways that might bring about deeper political and structural transformations toward a more ecologically responsive system of governance at the national and international levels. This requires "disciplined imagination," that is, drawing out a normative vision that has some points of engagement with emerging understandings and practices. Nonetheless, the role of imagination—thinking what "could be otherwise"—should not be discounted. As Vincent also points out, "We should also realise that to innovate in State theory is potentially to change the character of our social existence."[5]

This inquiry thus swims against a significant tide of green political theory that is mostly skeptical of, if not entirely hostile toward, the nation-state. Indeed, if a green posture toward the nation-state can be discerned from the broad tradition of green political thought, it is that the nation-state plays, at best, a contradictory role in environmental management in facilitating both environmental destruction and environmental protection and, at worst, it is fundamentally ecocidal.[6] From eco-Marxists to ecofeminists and ecoanarchists, there are few green political theorists who are prepared to defend the nation-state as an institution that is able to play, on balance, a positive role in securing sustainable livelihoods and ecosystem integrity.[7] It is now a trite observation that neither environmental problems nor environmentalists respect national borders and the principle of state sovereignty, which assumes that states ought to possess and be able to exercise more or less exclusive control of what goes on within their territories. Indeed, those interested in *global* political ecology are increasingly rejecting the "statist frame" through which international relations and world politics have been traditionally understood, preferring to understand states as but one set of actors and/or institutions among myriad actors and institutions on the global scene that are implicated in ecological destruction.[8] Thus many global political ecologists tend not only to be skeptical of states, they are also increasingly sceptical of state-centric analyses of world politics, in general, and global environmental degradation, in particular.[9] Taken together, the analyses of green theorists and activists seem to point toward the need for alternative forms of political identity, authority, and governance that break with the traditional statist model of exclusive territorial rule.

While acknowledging the basis for this antipathy toward the nation-state, and the limitations of state-centric analyses of global ecological degradation, I seek to draw attention to the positive role that states have played, and might increasingly play, in global and domestic politics. Writing more than twenty years ago, Hedley Bull (a proto-constructivist and leading writer in the English school) outlined the state's positive role in world affairs, and his arguments continue to provide a powerful challenge to those who somehow seek to "get beyond the state," as if such a move would provide a more lasting solution to the threat of armed conflict or nuclear war, social and economic injustice, or environmental degradation.[10] As Bull argued, given that the state is here to stay whether we like it or not, then the call to get "beyond the state is a counsel of despair, at all events if it means that we have to begin by abolishing or subverting the state, rather than that there is a need to build upon it."[11]

In any event, rejecting the "statist frame" of world politics ought not prohibit an inquiry into the emancipatory potential of the state as a crucial "node" in any future network of global ecological governance. This is especially so, given that one can expect states to persist as major sites of social and political power for at least the foreseeable future and that any green transformations of the present political order will, short of revolution, necessarily be state-dependent. Thus, like it or not, those concerned about ecological destruction must contend with existing institutions and, where possible, seek to "rebuild the ship while still at sea." And if states are so implicated in ecological destruction, then an inquiry into the potential for their transformation or even their modest reform into something that is at least more conducive to ecological sustainability would seem to be compelling.

Of course, it would be unhelpful to become singularly fixated on the redesign of the state at the expense of other institutions of governance.

States are not the only institutions that limit, condition, shape, and direct political power, and it is necessary to keep in view the broader spectrum of formal and informal institutions of governance (e.g., local, national, regional, and international) that are implicated in global environmental change. Nonetheless, while the state constitutes only one modality of political power, it is an especially significant one because of its historical claims to exclusive rule over territory and peoples—as expressed in the principle of state sovereignty. As Gianfranco Poggi explains, the political power concentrated in the state "is a momentous, pervasive, critical phenomenon. Together with other forms of social power, it constitutes an indispensable medium for constructing and shaping larger social realities, for establishing, shaping and maintaining all broader and more durable collectivities."[12] States play, in varying degrees, significant roles in structuring life chances, in distributing wealth, privilege, information, and risks, in upholding civil and political rights, and in securing private property rights and providing the legal/regulatory framework for capitalism. Every one of these dimensions of state activity has, for good or ill, a significant bearing on the global environmental crisis. Given that the green political project is one that demands far-reaching changes to both economies and societies, it is difficult to imagine how such changes might occur on the kind of scale that is needed without the active support of states. While it is often observed that states are too big to deal with local ecological problems and too small to deal with global ones, the state nonetheless holds, as Lennart Lundqvist puts it, "a unique position in the constitutive hierarchy from individuals through villages, regions and nations all the way to global organizations. The state is *inclusive* of lower political and administrative levels, and *exclusive* in speaking for its whole territory and population in relation to the outside world."[13] In short, it seems to me inconceivable to advance ecological emancipation without also engaging with and seeking to transform state power.

Of course, not all states are democratic states, and the green movement has long been wary of the coercive powers that all states reputedly enjoy. Coercion (and not democracy) is also central to Max Weber's classic sociological understanding of the state as "a human community that (successfully) claims the *monopoly of the legitimate use of physical force* within a given territory."[14] Weber believed that the state could not be defined sociologically in terms of its *ends*, only formally as an organization in terms of the particular *means* that are peculiar to it.[15] Moreover his concept of legitimacy was merely concerned with whether rules were accepted by subjects as valid (for whatever reason); he did not offer a normative theory as to the circumstances when particular rules *ought* to be accepted or whether beliefs about the validity of rules were justified. Legitimacy was a contingent fact, and in view of his understanding of politics as a struggle for power in the context of an increasingly disenchanted world, likely to become an increasingly unstable achievement.[16]

In contrast to Weber, my approach to the state is explicitly normative and explicitly concerned with the purpose of states, and the democratic basis of their legitimacy. It focuses on the limitations of liberal normative theories of the state (and associated ideals of a just constitutional arrangement), and it proposes instead an alternative green theory that seeks to redress the deficiencies in liberal theory. Nor is my account as bleak as Weber's. The fact that states possess a monopoly of control over the means of coercion is a most serious matter, but it does not necessarily imply that they must have frequent recourse to that power. In any event, whether the use of the state's coercive powers is to be deplored or welcomed turns on the purposes for which that power is exercised, the manner in which it is exercised, and whether it is managed in public, transparent, and accountable ways—a judgment that must be made against a background of changing problems, practices, and understandings. The coercive arm of the state can be used to "bust" political demonstrations and invade privacy. It can also be used to prevent human rights abuses, curb the excesses of corporate power, and protect the environment.

In short, although the political autonomy of states is widely believed to be in decline, there are still few social institution that can match the same degree of capacity and potential legitimacy that states have to redirect societies and economies along more ecologically sustainable lines to address ecological problems such as global warming and pollution, the buildup of toxic and nuclear wastes and the rapid erosion of the earth's biodiversity. States—particularly when they act collectively—have the capacity to curb the socially and ecologically harmful consequences of capitalism. They are also more amenable to democratization than corporations, notwithstanding the ascendancy of the neoliberal state in the increasingly competitive global economy. There are therefore many good reasons why green political theorists need to think not only critically but also constructively about the state and the state system. While the state is certainly not "healthy" at the present historical juncture, in this book I nonetheless join Poggi by offering "a timid two cheers for the old beast," at least as a potentially more significant ally in the green cause.[17]

1.2 Aims and Method: Critical Political Ecology

The perspective that I call critical political ecology is one that builds on the broad tradition of critical theory, giving it a distinctly green inflection.[18] With roots in the philosophies of Kant, Hegel, and Marx, and the Frankfurt School of Social Research, critical theory, as Richard Devetak has succinctly explained, is today recognized "as the emblem of a philosophy which questions modern social and political life through a method of immanent critique."[19] Andrew Linklater has called this method "praxeology," which he explains as the practice of critically reflecting on and harnessing those moral resources within existing social arrangements that might enable new forms of community with higher states of freedom.[20] Typically this entails critically questioning the values and norms that are internal rather that external to existing understandings and practices; exposing unfulfilled emancipatory promises and opportunities; unmasking tensions, contradictions, and hidden forms of coercion within and/or between ideas and practices; and exploring what historically possible changes in thought and practice might permit, facilitate, and/or enhance emancipation and enlightenment. This is the sense in which Max Horkheimer had asserted that "[a]gain and again in history, ideas have cast off their swaddling clothes and struck out against the social systems that bore them."[21]

Critical theory seeks a level of social understanding that transcends the *unreflective* understanding of historical agents, thereby also transcending the behaviorist program of social research, whose aim is merely to discern the meaning of the agents' self-understanding, taken at face value, by an "impartial social scientist." Unlike liberals, critical theorists do not take agents' preferences, needs, wants, or explicit avowals of belief as self-evident or as necessarily forming a coherent unity. The *critical* orientation of critical theory, with its abiding concern to uncover structures of domination, necessarily entails a refusal to accept the status quo or what passes for common sense. However, the point is not to discover what is really true or false but rather what is found to be more rational, by which I mean *reflectively acceptable* by social actors.

Critical theory's approach to critical reflection is thus based on a post-positivist, social constructivist theory of knowledge. This is what brings together critical and constructivists theorists, despite differences in their areas of focus (e.g., the former are typically more preoccupied with metatheoretical questions, whereas the latter more typically engage in empirical research into the role of norms and the social construction of identities). As Richard Price and Christian Reus-Smit point out, constructivism builds on critical theory's critique of positivism and "value-neutral" theorizing as well as its critique of rational choice theories of human nature. Claims that there is an objective reality are interpreted as always and unavoidably evaluative, historically contingent, and filtered through different social frames and social standpoints.[22] In short, all knowledge reflects particular social purposes, values, interests, and story lines, and this insight extends as much to our understanding of the so-called natural world as it does to the social world.[23] In view of the significant commonalities between critical theory and constructivism, I will enlist the composite term "critical constructivism" throughout this inquiry as an alternative to liberal and rational actor models of social choice.

The critical political ecology perspective that I seek to develop builds on the insights of critical constructivism by extending the project of emancipation to include both the human and the nonhuman world. Indeed, this had already been a preoccupation of the classical Frankfurt school, although succeeding generations of critical theorists have not continued this focus in any systematic way.[24] Critical political ecology seeks to rehabilitate the classical Frankfurt school's preoccupation with the links between the domination of human and nonhuman nature, while also building on more recent kindred developments in radical environmental philosophy and green political thought.[25] Whereas critical theory's quest for emancipation and enlightenment is a project that seek to question exclusionary practices and extend the boundaries of the moral community to include excluded and subaltern groups, critical political ecology may be understood as expanding this quest by extending the understanding and boundaries of the moral community to include not only the community of humankind but also the broader biotic community (in which human communities are embedded).

A central insight of ecofeminism and the environmental justice movement is that the domination of nature is a complex phenomenon that has been managed and mediated by privileged social classes and impersonal social and economic systems that have systematically brought benefits to some humans at the expense of others. The result is that certain privileged social classes, social groups, and nations have achieved what Mary Mellor, building on the work of Martin O'Connor, has called a "parasitical transcendence" from human and nonhuman communities.[26] In effect, a minority of the human race has been able to deny ecological and social responsibility and transcend biological embodiment and ecological limits (i.e., achieve greater physical resources, more time, and more space) *at the expense of others*, that is, by exploiting, excluding, marginalizing, and depriving human *and* nonhuman others. Val Plumwood has encapsulated this problem in the idea of remoteness. That is, privileged social classes have been able to remain remote (spatially, temporarily, epistemologically, and technologically) from most of the ecological consequences of their decisions in ways that perpetuate ecological irrationality and environmental injustice.[27]

Ultimately the vantage point of critical political ecology, when applied to environmental politics and the state, is one that seeks to locate and incorporate the demand for social and environmental justice in the broader context of the demand for communicative justice. By environmental justice I mean, first, a fair distribution of the benefits and risks of social cooperation and, second, the minimization of those risks in relation to an expanded moral community. By communicative justice I mean a fair/free communicative context in which wealth and risk production and distribution decisions takes place in ways that are reflectively acceptable by *all* "differently situated others" (or their representatives) who may be affected.

1.3 Working toward the Green State: A Provisional Starting Point

The popular philosophy of the green movement has a well-recognized position. In matters of institutional design and its programmatic defense of the principles of decentralization, grassroots democracy, and nonviolence, its motto is "Think globally, act locally." However, what is striking is that these principles often sit considerably at odds with the day-to-day campaign demands of environmental activists, organizations, and green parties for "more and better" state regulation of economic and social practices in order to secure the protection of the environment.[28] Indeed, the same has been said of new social movements in general, which tend, on the one hand, to "subscribe to antistatist slogans and the fundamentalist critique of the state's 'monopoly of force,' while, on the other hand, they propose large doses of state resources (both fiscal and repressive) to be made available to the causes of desired social change."[29] Should we regard this a fundamental contradiction in green thought and practice or, as Matthew Paterson suggests, merely a necessary ambiguity of green politics?[30] Much depends on whether the greens' strategic associations and negotiations with the state undermine or reinforce their vision of what a good state might look like, and whether the vision is defensible. Either way it seems clear that the green movement *needs* the state (in some if not all respects) if it is to move closer toward its vision of a socially just and ecologically sustainable society. But would the state be enlisted merely *instrumentally* in the social and political struggle to achieve green goals and/or would it be regarded as some kind of *embodiment* of the public virtue or democratically determined public values?

A good place to start is to explore what sort of state would emerge if the green movement's programmatic demands for more environmental regulation were successfully and fully pursued over a sustained period of time. In short, what conception of politics, public life, and the state lies behind the green demands made of the state, and how might this be practically embodied more explicitly in the formal constitutional structure and informal political culture of states?

There seem to be two basic interrelated ideals about the state implicit in the demands for environmental regulation and justice. The first is a plea for a strong or effective state. The second, which legitimizes this disciplinary face of the state, is a plea for a good state, in the sense of an ethical and democratically responsible/responsive state that upholds public interests and values, and acts as a vehicle for environmental justice rather than self-serving power.

That the state should be "strong" or effective arises from the need to facilitate environmental restoration, regulate, and in some cases proscribe a wide range of environmentally and socially damaging activities. Essentially this call upon the state seeks the deployment of the regulatory and fiscal steering mechanisms of the state to ensure that the economy and society respect the integrity of the ecosystems in which they are embedded. The state is enlisted because it is the social institution with the greatest capacity to discipline investors, producers, and consumers. (Markets—as social institutions—have a more limited capacity to turn green, and they are not amenable to the same degree of citizen control; at best, they are responsive to consumer sovereignty rather than to political sovereignty or a politically constituted public.) The state also has the capacity to redistribute resources and otherwise influence life opportunities to ensure that the move toward a more sustainable society is not a socially regressive one—a very real prospect if environmental goals are not properly integrated with social justice goals. This state capacity arises precisely because it enjoys a (virtual) monopoly of the means of *legitimate* coercion and is therefore the final adjudicator and guarantor of positive law. In short, the appeal of the state is that it stands as the overarching political and legal authority within modern plural societies.

This appeal to the "strong" or effective state should not be understood as an entirely instrumental appeal; otherwise, there would be no reason, in principle, for environmentalists not to hire private mercenaries to discipline society along more ecologically sustainable lines, assuming that the necessary resources can be mustered. That the state should also be "good" arises from the understanding that the state is (potentially) the most *legitimate*, and not just the most powerful, social institution to assume the role of "public ecological trustee," protecting genuinely *public* goods such as life-support services, public amenity, public transport, and biodiversity. Such a normative posture toward the state harks back to the European idea of the state as the embodiment of reason, ethics, and the collective good. In this respect this view is reminiscent of the civic republican tradition insofar as the laws of the democratic state are enlisted to constitute (as distinct from merely restrict) the ecological freedom of all citizens. As Jürgen Habermas has argued, the law in democratic societies has a dual character in that it provides the substantive and formal rules to stabilize, integrate, and regulate society as well as the democratic procedural requirements to ensure the legitimacy of such regulations.[31] It is precisely these democratic procedural requirements that convert the state's coercive power into *legitimate* coercive power.

Finally, there is the hope in green demands upon the state that it would not only act as a good ecological trustee over its own people and territory but also as a good international citizen in the society of states. It is implicit that the green state actively promote collective action in defense of environmental protection and environmental justice while also taking responsibility (both unilaterally and multilaterally) to avoid the displacement of social and ecological costs beyond its own territory and into the future.

In these times of increasing globalization and continuing state rivalry there are likely to be many sceptical responses to this normative vision of the state, from both within and beyond the green movement. Doubtless there are other implicit visions of the state that may be drawn out of any particular set of environmental public policies. Nonetheless, I will take this normative ideal as a provisional starting point, as something that is worth seriously pursuing. The rest of this book is concerned to explore criticisms and challenges to this ideal and to suggest how it might be fleshed out, and to what extent it might be necessary to reconstruct it in response to such criticisms and contemporary exigencies. Consistent with the method of what I now call critical political ecology, the path I have sought to tread in the following chapters is one that seeks to navigate between undisciplined political imagination and pessimistic resignation to the status quo.

1.4 Three Core Challenges

Since questions of democracy and legitimacy are intimately tied up with questions of political autonomy and functional capacity, it is necessary to answer those critics who might reasonably argue that the very notion of a "green democratic state" is merely wishful in the sense that it faces insuperable challenges. I have singled out what I take to be the three core challenges or "hesitations" to the prospect of greening the state and the state system. These core challenges are:

1. The anarchic character of the system of sovereign states. The problem is understood as structuring a dynamic of selfish and rivalrous behavior among states that results in the all-too-familiar "tragedy of the commons."
2. The promotion of capitalist accumulation. The way in which the state is inextricably bound up with, and fundamentally compromised by, globalization is also a key driver of ecological destruction. States are now actively promoting economic globalization in ways that further undermine their own political autonomy and steering capacity.
3. The "democratic deficits" of the liberal democratic state. The liberal state is regarded by many green political theorists as suffering too many democratic deficits to be able to respond to ecological problems in a reflexive and concerted manner. This critique is directed not only to the instrumental rationality of the "administrative state" but also to the *liberal* character of its democratic regulative ideals, which are seen as inhibiting the protection of public goods such as the environment.

Together, these different challenges capture what I take to be the most significant and enduring obstacles in the way of enlisting and reforming the state as a site and agent of ecological emancipation. They suggest that the prospects for the development of more ecologically responsive states are bleak and possibly hopeless. Any critical reconstruction of the normative vision of the green democratic state outlined above must therefore wrestle with these challenges and explore how they may interact in mutually reinforcing or countervailing ways. In chapters 2, 3, and 4, I address each of these three challenges respectively.

The overall argument that I offer is that it is too hasty to assume that the social structures of international anarchy, global capitalism, and the liberal democratic state are necessarily anti-ecological and mutually reinforcing, or that they foreclose the possibility of any progressive transformation of states as governance structures. The key to such transformation lies in deepening the democratic accountability and responsiveness of states to their citizens' environmental concerns while also extending democratic accountability to the environmental concerns of transnational civil society, intergovernmental organizations and the society of states in general. By these means, the anti-ecological behavioral dynamics that are generated by the social structures of international anarchy, global capitalism and administrative hierarchy can be reversed. One does not have to search very far to find historical examples of how environmentally destructive dynamics can be qualified, restrained, or otherwise moderated by state and nonstate agents "acting back" upon social structures. Here I single out three mutually informing developments that have served to moderate and, in some cases, transform the respective "logics" of international anarchy, capitalism, and administrative hierarchy:

1. The rise of environmental multilateralism, including environmental treaties, declarations, and international environmental standards.
2. The emergence of sustainable development and "ecological modernization" as competitive strategies of corporations *and* states.
3. The emergence of environmental advocacy within civil society and of new democratic discursive designs within the administrative state, including community "right to know" legislation, community environmental monitoring and reporting, third-party litigation rights, environmental and technology impact assessment, statutory policy advisory committees, citizens' juries, consensus conferences, and public inquiries.

In circumstances where these three developments can be found to operate in *mutually reinforcing ways*, it is possible to glimpse a possible trajectory of development that moves away from "organized ecological irresponsibility" (to adapt Ulrich Beck's phrase) to more ecologically responsible modes of state governance in the areas of economic development, social policy, security, and diplomacy.[32] However, it is a central argument of this book that the likelihood of this trajectory ever being realized is crucially dependent on the degree to which states can be made more democratically accountable in terms of a distinctly green rather than liberal conception of democratic state governance.

Accordingly, in chapter 5, I outline an ambit claim for ecological democracy as an alternative to liberal democracy and then explore its scope and character in the nation-state. I defend ecological democracy as more conducive than liberal democracy to reflexive societal learning, as it is better placed to minimize ecological risks and avoid their unfair displacement onto innocent third parties in space and time.

In chapters 6 and 7, I examine how far the ambit claim for ecological democracy might be embodied in the constitutional framework of the green democratic state, and how and to what extent it might be practically realized, both domestically and transnationally. In both chapters I work toward a distinctly green theory of the democratic state by distinguishing it from liberal as well as alternative civic republican accounts while situating it in the context of recent critical theories of the state, civil society, and the "green public sphere." I develop this normative theory out of a critical review of the most influential rival theory to both liberalism and republicanism, notably the discourse theory of law, democracy, and the state offered by critical theory's most influential contemporary scholar—Jürgen Habermas.

In chapter 6, I show how the green democratic state can be defended as being more legitimate than the liberal democratic state. I show how it seeks to both deepen and extend democracy in ways that are more sensitive to the highly pluralized context of today's societies confronting complex ecological problems in an increasingly borderless world. However, the project of building the green state can never be finalized. Rather, it is a dynamic and ongoing process of extending citizenship rights and securing more inclusive forms of political community. The flourishing green public sphere is crucial to this process, and I suggest how the mutually dependent relationship between the green democratic state and the green public sphere might be held in creative balance.

In chapter 7, I explore the transboundary dimensions of ecological democracy and defend what I call the transnational green democratic state as an alternative to both civic republican and global liberal cosmopolitan accounts of democracy. I argue that the cosmopolitan democratic principle, which also underpins the ambit claim for ecological democracy, that all those potentially affected by proposed norms/risks should be entitled to participate in the making of decisions, should not form the basis for deciding what should be the *primary* unit of governance. However, I show how "affectedness" may come into play in the development of supplementary structures of rule that create transboundary rights of ecological citizenship. I argue that this supplementary structure of rule should be developed by multilateral negotiations. Such an approach is defended as both more desirable and more feasible than the development of cosmopolitan democratic governance at the global level.

Finally, in chapter 8, I draw out some of the significant shifts in global discourses on environment, development, security, and intervention over the past four decades. In particular, I show how gradual changes in shared understandings of the development rights and environmental responsibilities of states have given rise to "green evolutions in sovereignty." I also explore how this trajectory might be furthered by a "negative sovereignty discourse" that argues that environmental harm is an unwarranted form of intervention in the territory and affairs of states. I end with a discussion on how the existing principle of state responsibility for environmental harm could develop into a more radical principle that might more effectively protect ecosystems and environmental victims while also extending the role and rationale of states to that of environmental custodians.

2

The State and Global Anarchy

2.1 Environmental *Realpolitiks* and the Tragedy of the Commons

In a recent critical assessment of the prospects for a green democratic state, Michael Saward has asked: "Could it be that the contemporary state is simply not the type of entity which is capable of *systematically* prioritizing the achievement of sustainability?"[1] Historically the defense of state territory, military success, and the exploitation of natural resources and the environment for the purposes of national economic development and national security have been widely understood as the overriding imperatives of all states and constitutive of the state's very form.[2] Indeed, the exploitation of natural resources within the territory (energy resources, timber, minerals, etc.) has sometimes been justified as a nation-building exercise or intimately linked with national security. According to this line of argument, as entertained by Saward,

> environmental protection and preservation would seem to run counter to the main imperatives constituting states: the need to secure economic stability and growth, the need to keep social order (primarily through welfare states in our era), and "staying afloat in a hostile world," involving military and other security imperatives.[3]

The idea that states are preoccupied with "staying afloat in a hostile world," and the implication that security imperatives are fundamental and overriding, are both hallmarks of the dominant realist perspective in International Relations theory. In this chapter, I critically explore the structural realist view that there are systemic pressures that set limits to more enlightened ecological governance by states and that arise by virtue of their status as *units in a state system*. My concern is not to reject this approach entirely but to highlight its limitations by placing it alongside two alternative readings of the anarchic international order—neoliberal institutionalism and critical constructivism—that suggest that the prospects for enlightened environmental governance are considerably brighter than neorealists would have us believe. I show that critical constructivism, in particular, is able to point to the changing practice of multilateralism, which carries the potential to broaden the roles and identities of states to include that of ecological steward, replacing the traditional role of environmental exploiter.

It is often noted by political ecologists that "the earth's political geography bears no resemblance to its appearance from space—a solitary blue planet, with a single ocean and seven large land masses."[4] Dividing up the earth in terms of invisible lines called political borders appears arbitrary from an ecological point of view. Moreover one might well be skeptical about the possibility of a single, complex, and highly integrated ecosystem, namely the biosphere and all its interlinked parts, being managed on an ecologically sustainable basis within the constraints of a political system made up of around 190 states, each jealously claiming the sovereign authority within their territory.[5] In short, there seem good reasons to believe that the system of sovereign states is inimical to the emergence of green states of the kind outlined in chapter 1.

Although my ultimate concern is a normative one—to reconstruct the rationale, functions, and democratic procedures of states along more ecologically sensitive lines—I seek to pursue this concern by means of a critical encounter with existing understandings of the role and purposes of the state as a member of a larger system or society. It is therefore necessary to respond to the realist perspective not only because it has dominated the study of International Relations (and the practice of statecraft) since the beginning of the cold war but also because it represents the most pessimistic assessment of the prospects of green democratic states emerging in the current order. Critical theorists are not exactly optimists either, but they do at least countenance the possibility of green or greener states emerging out of the contradictions generated by global capitalism and the political mobilization of new social movements. In contrast, realists, who privilege "the relations of destruction" over "the relations of production," see the state system as self-perpetuating in ways that foreclose enlightened ecological governance. The problem with realism's restricted focus of inquiry is that it is dismissive of any preoccupation with an appropriate ethics of international relations and is concerned only to develop an "objective" science of international relations. However, I do not thereby ignore the "reality" of power struggles nor the unequal distribution of military and economic power among states. Indeed, one should remain wary of imputing or exaggerating the existence of common interests among nations.[6]

That states should be so preoccupied with security issues is understood by realists to arise from their location in an anarchic state system, which is understood as an essentially Hobbesian world made up of asocial, strategic state actors who are fearful, mistrustful, and constantly competing for scarce resources for the purposes of self-preservation or expansion. Moreover success in either of these pursuits is understood to be a function of the different material capabilities of states, which boils down to their military strength and economic power (seen as mutually reinforcing, for the most part). For traditional realist scholars such as Hans Morgenthau, states are engaged in a constant struggle for power: war is a constant threat and peace is only attainable where there is a "balance of power" to stabilize relations among states.[7] For neorealists such as Kenneth Waltz, the rivalrous strategic behavior of states is something that is generated by the very anarchic structure of the state system despite the putative "higher rationality" of alternative arrangements and arguments, green or otherwise. On this pessimistic view, beyond the formation of strategic alliances, the prospects for interstate cooperation on environmental matters would appear dim. Since security and economic interests are the only serious matters of high politics, environmental protection is forever condemned to the periphery of international relations, including the discipline of International Relations.[8] As the British former Prime Minister Margaret Thatcher once put it in her own inimitable way when speaking about the Falklands war: "It is exciting to have a real crisis on your hands when you have spent half your political life dealing with humdrum issues like the environment."[9] Those pushing or hoping for more enlightened global environmental governance are therefore liable to be condemned by realists for engaging in wishful and possibly dangerous thinking, on a par with the idealists of the inter-war period.[10]

And since the parameters of domestic politics are constrained by overriding systemic pressures, we can also expect that environmental protection will be relegated to the periphery of states' domestic politics as well, and increasingly so as the competitive pressures of economic globalization intensify.

Neorealism offers not only a fundamental explanation for the "tragedy of the commons" writ large but also an explanation for environmental degradation within state territories.[11] Put bluntly, it is not in the "interests" of states to take concerted action to protect the global commons, the biosphere, or even the ecological integrity of their own territory ahead of more "fundamental" security and economic goals. States are not *obliged* to enter into cooperative regimes to protect the global environment, and if they do, they are free to implement and manage regimes in ways that protect their own strategic "interests." As rational egoists, they have no incentive to take unilateral action to protect the environment whenever this might create costs or disadvantages relative to other states. State rivalry, the temptation to "free ride" and the enforcement problems associated with attempts to protect collective goods, the interminable conflicts over apportioning the burden and costs of environmental reforms among nations (particularly between developed and developing states), and the jealous protection of sovereign territorial rights are all seen to conspire to make protection of the global and domestic environment a serious uphill battle. In short, environment problems are simply not considered important enough to dislodge the more basic, and base, state interests of survival/security and economic advancement. Moreover the structural imperatives created by international anarchy are understood to leave no or little room for any diversity of state responses to domestic and collective problems, since all states are, to borrow Kenneth Waltz's phrase, "unit like" and therefore respond in the same way to systemic pressures.

Now, if there is one development that might dent this realist analysis of the peripheral nature of environmental problems, it is when ecological problems begin to pose a direct threat to the fundamental security or economic interests of states. Under these circumstances we might expect environmental problems to be regarded by states as matters of high rather than low politics, especially when they threaten the very existence or territorial integrity of states. Such a situation is no more graphically illustrated than in the case of low-lying island states that face virtual extinction if the predicted consequences of global warming are borne out.[12] Yet even in these extreme scenarios realists offer the dismal warning that we cannot assume that interstate cooperation will occur or, if it does, that it would be successful in addressing collective ecological problems where ecological vulnerability, the costs of adjustment, and the economic and infrastructural capacity to adjust are unevenly distributed among the states (as is presently the case). This is because altruistic concern about the fate of the more vulnerable states is ruled out and selfish rivalry is played out in the pursuit of relative rather than absolute gains. Moreover, insofar as states have managed to create environmental agreements on collective problems, they have not been the product of free and fulsome deliberation but rather due to coercive and/or strategic bargaining that reflects narrowly conceived geopolitical interests. Yet, in an anarchic world order where "might is right" and notions of morality, ethical behavior, and the common good have no place, any agreement reached by states will always be vulnerable to shifts in the distribution of power.

Now, if we accept, for argument's sake, this narrow and pessimistic reading of the rationality of states as international actors, then we might also note, as Garret Hardin and many game theorists have, the obvious ecological irrationality of the dynamic that it describes. That is, the "rationality" of states in an anarchic system is such that they are unable to act collectively to address collective ecological problems, even when such problems threaten to undermine the very territorial integrity of some states. We saw this "security dilemma" played out in a nuclear arms race that only heightened rather than alleviated insecurity (although the chilling logical terminus of this race—mutually assured destruction (MAD)—ultimately acted as a perverse form of deterrence). We now see its ecological counterpart played out in the escalation of ecologically damaging military activities and economic development whose dire consequences are emerging as certain rather than merely potential future threats. In short, from the realist tradition, we can expect states to engage in the unrestrained exploitation of natural resources, species, and ecosystems. Of course, realists do not *advocate* this as an "environmental ethic." However, it is understood to be the orientation of the state flowing from realism's analysis, and since realists do not countenance other ways of thinking about the relationships between states and their environments, they render such an exploitative orientation toward the environment as natural and inevitable. Realists may, in effect, be accused of being complicit in the perpetuation of such exploitation.

A common criticism of neorealism is that in privileging structure over agency, it provides an essentially static understanding of the international order that cannot account for change at the level of unit or system nor even acknowledge the potential for such change. As Richard Ashley succinctly explains, "it denies the role of practice in the making and possible transgression of social order."[13] The interests, identities, and roles of states are "essentialized" (and assumed to remain fixed) since the anarchic structure of the state system is understood as overdetermined and therefore essentially unaffected by changes in domestic politics or by the emergence of national or transnational counterhegemomic protests and discourses. This understanding of the relationship between social agents and social structures stands in sharp contrast to that of critical constructivism, which seeks to comprehend historical change as the result of the changing relationships between structures *and* agents. Since structures are produced by the recurrent practices of agents, changes in the patterns of interaction among agents can produce changes in social structures. In principle, no social structure is immutable and beyond transformation, although many social structures can be very persistent, "sticky," and highly resistant to change. As Linklater explains, "whereas neo-realism aims to account for the reproduction of the system of states, critical theory endeavours to highlight the existence of counterhegemonic or countervailing tendencies which are invariably present within all social and political structures."[14]

The normative potential of such counterhegemonic tendencies is explored later in the chapter. For the moment it is necessary to develop a critical appreciation of why states have repeatedly transgressed realists' predictions in their diplomatic relationships and environmental multilateral arrangements. To clarify our task, the point here is not to demonstrate that the realist analysis is irrelevant to our normative task. Just because one can learn to think outside the realist frame does not mean that the security dilemma or the tragedy of the commons is thereby resolved. Realism still illuminates the behavioral dynamics of those states living with the constant threat of military conflict, and it is likely that some transboundary environmental problems are likely to be a source of increasing instability and conflict in the new millennium (especially in relation to scarce water resources). Moreover the pursuit of national security is something that should concern environmentalists generally, not only because it can sometimes lead to the suspension of, or encroachment upon, civil and political rights that are essential to the practice of ecological citizenship but also because it can be, among other things, environmentally destructive. Military training, weapon production, storage, disposal, and, above all, armed conflict have proved to be major causes of the most serious ecological degradation of the last century (especially nuclear, chemical, and biological testing and warfare). Exploring how old cycles of conflict might be broken in order to pursue a path of nuclear nonproliferation and interstate cooperation is one of the most urgent and significant steps toward global ecological integrity.[15] Avoiding the development of an authoritarian "security state" by ensuring greater public and environmental accountability of defense policy and defense activities should likewise be essential to any green public sphere and green democratic state. Nonetheless, it would be naïve for greens to believe that states can function without military and police forces. Once this is acknowledged, then from a democratic green perspective, everything should turn on whether those forces are deployed in legitimate ways to further legitimate ends—a point that takes us back to changing normative rationales of states as governing entities.

While one may concede that the neorealist understanding is by no means irrelevant to our understanding of global environmental degradation, for present purposes it is nonetheless heavily lopsided and reductionist. This arises from neorealism's limited, a priori assumptions about the character of the international order and the motivations and interests of states. These assumptions may have had resonance during the cold war era, but they are less pertinent in the contemporary world. Rivalrous state behavior is certainly implicated in a tragedy of the commons. However, I hope to show that the reasons for this tragedy are even more complex, and in many cases more mundane and varied, and the prospects for addressing this tragedy through the medium of states are brighter than realists might imagine.

There is now a developed critique of realism within the discipline of international relations from liberals as well as critical and constructivist theorists. These critics have pointed out that states are not solely preoccupied with power and physical and economic security; that they are not the only actors in world politics; that states do successfully cooperate for mutual benefit in many domains of mutual concern; that morality is not irrelevant to world politics; that multilateral institutions can influence the identities, interests, and purposes of states; and that the boundaries between international and domestic politics are by no means clear-cut.[16] Indeed, the idea that international institutions are maintained only by force, the threat of force, or strategic calculation is no less fanciful than the idea that they are maintained only by a moral consensus arrived at after free and fulsome deliberation. Taken together, these insights hold some promise for the greening of the state and the society of states, if not the emergence of a world of full-fledged green states, as I hope to show.

One reason why realists seem to ignore or marginalize these significant developments in environmental multilateralism is that they simplify, and therefore misconceive, the relationship between power and morality, and between material interests and ideas/culture, in world politics. While it is true that the different material capabilities of states have a significant bearing on the ability and motivation of states to manage their own environments and enter into cooperative agreements with other states to manage transboundary ecological problems, this is not the whole story. Simply ranking states in terms of their military and economic strength does not enable us to predict or understand the success or otherwise of, say, the International Criminal Court, the land mines convention, or the climate change negotiations (all of which have proceeded despite the noncooperation of the United States). It is not only brute power in the form of technologies (military hardware, machines, etc.) or brute economic strength measured in wealth but also different sets of *shared understandings* about who controls, owns, manages, and decides, and about social objectives, social obligations, and modes of accountability, that ultimately determine what happens. It is only by exploring these shared understandings, and the actors that support and contest them, that it is possible to develop an appreciation of how material power is understood and deployed, legitimately or otherwise.[17] From the perspective of critical political ecology, the question of shared understandings (and how they are reached, maintained, and/or transformed) is essential to understanding both continuity and change in any social order. Such an approach also directs attention to the social basis of legitimacy, in particular, to the ways in which changes in shared understandings about the meaning and purpose of social life can undermine and/or alter the basis of the legitimacy of particular social structures, including states.

Even if one were to accept the controversial realist assumption that states everywhere always have an overriding "interest" in security and economic development, these "interests" themselves have already undergone some degree of ecological reconceptualization, manifest in the new discourses of "ecological security," "sustainable development," or "ecological modernization." While these new discourses remain highly contested, it is nonetheless possible to discern from them at least the outlines of new ecological understandings concerning how states should secure their territory, develop their economies, and ensure the welfare of their peoples. Indeed, the emergence of these ecological discourses on security and development, some of which have found their way into domestic and foreign policies and multilateral environmental agreements, makes it already possible to talk of the modest greening of the rationale of states. As I show in chapter 8, the principle of state sovereignty is not a self-justifying norm but rather takes its meaning from the changing constitutive discourses that underpin it (those that determine the rules of intervention, the meaning and scope of self-determination, the meaning and scope of security, the right to develop, etc.).[18] To the extent to which the constitutive discourses of sovereignty take on an ecological dimension, it becomes possible to talk about the concomitant greening of sovereignty. While realism's mainstream rival—neoliberal institutionalism—has taken environmental multilateralism seriously, it does not offer a framework of understanding that is interested in, or capable of detecting, changes at this deeper structural level.

2.2 Neoliberalism, Environmental Regimes, and the Limits of Problem Solving

The fact that international law is observed by most states most of the time, and that environmental cooperation between states does routinely occur (however suboptimal that may often be), is not something that the Hobbesian, state-centric framework of realism can satisfactorily explain. The rapid proliferation of new multilateral environmental treaties, declarations, and strategies and international environmental organizations in recent decades—prompted in no small way by the proliferation of environmental NGOs—cannot be simply dismissed as exceptions that prove the neorealist rule nor as mere reflections of the changing balance of power among nations. Ironically it has been the further development of rational choice theory, particularly more sophisticated approaches to game theory, that has gone some way toward explaining why cooperation between states often turns out to be more likely than defection in situations of complex interdependence, and why it is that the prospect of absolute gains, rather than relative gains, can be sufficient to secure cooperation between states.[19] Neoliberal institutionalists have made a major contribution to our understanding of environmental regimes in directing attention to the ways in which different institutional settings can affect the motivation of states to cooperate to address collective environmental problems. A regime is generally understood to refer to a set of "implicit or explicit principles, norms, rules, and decision-making procedures around which actors' expectations converge in a given area of international relations."[20] This understanding can stretch to encompass treaties as well less formal agreements such as declarations and strategies (provided the necessary "convergence of expectations" can be found). Indeed, the bulk of academic work on environmental multilateralism is now conducted within a neoliberal "interest-based" institutional framework of explanation rather than a realist or neorealist "power-based" framework, although the differences between these two approaches are sometimes blurred.[21] Although neoliberal institutionalists do not ignore the distribution of power among states, they have shown that environmental regimes are rarely a simple reflection of this distribution and that understanding the broader constellation of interests associated with different regimes provides a better clue to understanding regime effectiveness. Borrowing heavily from economic theories of institutions (with their focus on information and transaction costs), neoliberal institutionalists have challenged realism's pessimism about the prospects of enlightened global ecological governance by showing that environmental regimes can, under appropriate circumstances, be both effective and robust in avoiding collectively suboptimal outcomes.[22] Thus institutions matter in the sense that they provide a means of coordinating and harmonizing interstate relations, thereby providing certainty and a framework of action that ultimately serves the sovereign rights of states by enabling rather than undermining self-preservation and material interest maximization. As Ken Conca has recently put it, for the realist, state sovereignty is the problem; it explains the failure of environmental governance. For the neoliberal, state sovereignty is the solution, at least in the sense that the emergence of multilateral institutions for environmental protection serve to maintain state capabilities and expand the menu of choices available to states.[23]

A basic difference between the neorealist and neoliberal approaches is that neoliberals understand international society in essentially Lockean rather than Hobbesian terms. Within this context states are posited as rational egoists engaged in instrumental calculations, and any cooperative agreements that are reached represent bargains that provide a better set of payoffs than alternative self-help arrangements. However, for neoliberals, international society is nonetheless a rather thin one in that any common good arising from such bargains is nothing more than the aggregation of the satisfactions of utility-maximizing states. Nonetheless, regimes—by enmeshing states in reciprocal rights and responsibilities—do provide a reflection of a rudimentary moral community based on a certain degree of trust and reciprocal recognition by the individual members of that community. It is precisely this trust and mutual recognition that enables international governance to take place despite the absence of a central international government.

For those of us interested in exploring the prospects for the development of green or at least greener democratic states, neoliberal institutionalism provides a more optimistic assessment than realism. Through their detailed and comparative studies of environmental treaties, in particular, neoliberals have offered a range of recommendations and insights as to how to manage complex interdependence by removing uncertainty and improving environmental cooperation in a system of sovereign states lacking a central authority. Environmental protection may still not be regarded as a fundamental rationale of states, but it has clearly emerged, in varying degrees, as a subsidiary purpose of states, evidenced in no small way by the proliferation of multilateral environmental treaties, declarations, and action plans since the time of the first United Nations Conference on the Human Environment in Stockholm in 1972.

However, environmental multilateralism has not, on the neoliberal institutionalist understanding at least, altered the basic structure of territorial rule in any fundamental respect. Indeed, the prospects for such a transformation are not even on the radar screens of neoliberals because, like realists, they take the state system, and state interests and identities, as unproblematic background immutables. Under these circumstances environmental multilateralism has ultimately served to shore up state sovereignty rather than fundamentally challenge it. Entering into multilateral arrangements that restrict what states can do is not an abrogation of sovereignty; rather, it as a voluntary *exercise* of sovereignty based on the principle of liberal contractualism. As I show in the following chapter, this stands in stark contrast to economic neoliberals, neo-Marxists, and global political ecologists, who claim, for different reasons, that multilateralism and economic globalization have conspired to *undermine* state sovereignty.

While neoliberal institutionalists are optimistic relative to neorealists, from the perspective of critical theory, this optimism must be understood in a narrow "problem-solving" way in Robert Cox's sense of the term. As Cox explains, "critical theory can be a guide to strategic action for bringing about an alternative order, whereas problem-solving theory is a guide to tactical actions which, intended or unintended, sustain the existing order."[24] On this view, environmental multilateralism can, at best, provide a means of ameliorating certain common and transboundary environmental problems. However, its issue by issue, problem-solving focus serves to bracket both the constitutional structure of international society along with the constitutive discourses of sovereignty that sustain it. Peter Haas, Robert Keohane, and Mark Levy, three leading advocates of neoliberal institutionalism, are quite explicit on this point: "We ask whether institutions can help retard the rate of environmental decline, even if they fail to confront the underlying causes of such decline. We are pragmatists."[25]

Moreover this pragmatic assessment of the prospects of more enlightened environmental governance by neoliberals is one that generally takes as given, rather than critically inquires into, the interests and identities of states. This is because states are assumed to be rational actors that undertake purely instrumental assessments of the costs and benefits of multilateral cooperation relative to noncooperation. To be sure, recent work by regime theorists has added some supplementary "interpretive insights" to this interest-based analysis by exploring the role of cognitive structures, "epistemic communities," transnational issue networks, and the more general question of social learning.[26] This has meant that ideas and norms are sometimes credited by neoliberals as having some influence on state behavior beyond the effects of material capabilities, interests, and institutions. However, ideas and norms tend to be brought in as ad hoc arguments emerging from outside the neoliberal framework when power, interests, and institutions cannot fully account for state behavior in particular contexts.[27] In short, they do not challenge the basic neoliberal understanding of regimes as an essentially *functional* response to collective action problems. As Alexander Wendt explains, a perspective that treats ideas and norms as merely "intervening or superstructural variables will always be vulnerable to the charge that they are derived from theories that emphasise the base variables of power and interest, merely mopping up unexplained variance."[28]

Thus from a neoliberal institutionalist perspective, unless environmental proposals can pass through the filter of a utilitarian calculation from the narrowly conceived interests of states, they are unlikely to become the subject of multilateral agreements. This would seem to rule out noninstrumental, moral arguments for the protection of the environment that do not converge with the material interests of states. At best, then, we can expect ongoing environmental multilateralism to foster the emergence of more pale green states whose material interests and identities as calculating, strategic actors remain unaltered.

Yet just as realists have no satisfactory explanation for the rise of environmental multilateralism, neoliberal institutionalists have no satisfactory explanation for those multilateral environmental regimes that cannot be easily reduced to the instrumental and material calculations of states. For example, there are many environmental regimes that reflect a strong preservationist or protective rather than merely "wise use" or resource conservationist perspective, such as those dealing with the transboundary movement of hazardous wastes, whaling, the protection of the Antarctic from mining, the restriction of trade in endangered species, and the protection of the ozone layer. Now it is often possible to recast preservationist arguments in self-interested, instrumental terms, just as it is possible to recast human rights norms in instrumental terms, but this typically does violence to the arguments, identities, and intentions of the state and nonstate advocacy networks and coalitions that raise and manage to more or less discursively defend such claims in multilateral negotiations. Moreover, as I argue in more detail below, the more the neoliberal notion of self-interest is redrawn to accommodate new environmental pressures and values, the more are alternative explanations squeezed out and overall explanatory power lost.

Of course, it cannot be denied that most environmental regimes and negotiations are also based on a good deal of strategic bargaining and haggling over the distribution of benefits and burdens. However, neoliberals tend to neglect the fact that this bargaining also typically takes place in a *moral context*. This is well illustrated in the current climate change negotiations, where haggling over benefit and burden sharing has been intense. Yet such haggling remains framed and constrained by a set of environmental protection and environmental justice norms negotiated in the United Nations Framework Convention on Climate Change signed in 1992.[29] Accordingly, to reduce the climate change negotiations to a set of bargaining positions based on relative vulnerability, the capacity to adjust and the costs of adjustments facing the parties is to neglect the most basic of all issues, which is the normative purpose of the Convention (to protect the world's climate in an equitable manner).[30] It is certainly a major problem that the United States—the state with the largest share of greenhouse gas emissions—has withdrawn from the negotiations. However, the negotiations have nonetheless proceeded despite this withdrawal, and the Bush administration has attracted considerable condemnation not only from many parties to the negotiations but also from US civil society and global civil society. In any event, Bush's posture cannot simply be deduced from "objective interests" but rather must be understood in terms of the particular ideological proclivities of the new administration.

Like neorealism, neoliberal institutionalism provides an account of certain "modal responses" by states "to certain types of structural constraints or situational exigencies,"[31] but it too provides a lopsided analysis that provides at best a partial understanding of both the practices and prospects of environmental multilateralism for the greening of both states and the society of states.

2.3 Critical Constructivism and Social Learning

If there is a common motif in critical theories of the state, it is one that emphasizes the paradoxical or contradictory character of the modern state as a site of inclusion and exclusion, emancipation and oppression, or, in our case, environmental protection and exploitation.[32] This means that neorealists and neoliberal institutionalists are often *partly right* in drawing attention to the state's traditional record of environmental destruction and its more recent, but not entirely successful, efforts toward more ecologically rational utilitarian management. But the story does not end there, since it is also possible to find not insignificant examples of states acting either individually or collectively as "ecological trustees," upholding a wider range of ecological values in the service of environmental protection. While modern statecraft has certainly helped to generate many ecological problems, the increasing prevalence of global ecological problems is now increasingly challenging modern statecraft, leading to new and sometimes highly innovative multilateral and domestic responses.

Moreover, in exploring the uneven history of these contradictory postures, critical constructivists reject the idea that the state system can be studied in isolation, as if states were always able to withstand changes in other political and institutional domains. Rather, shifts in domestic politics, the transnational activity of NGOs (e.g., corporations, scientific communities, environmental organizations, and the media), and the emergence of new problems and discourses in national and global civil society can serve as major catalysts for changes at the international level, and vice versa. To understand these developments fully, it is necessary to bridge the increasingly unhelpful disciplinary division between international relations, domestic political studies of state/society relations, and the sociology of globalization. Moreover, understanding the potential for social transformation entails exploring structural contradictions and political resistance. As neo-Gramscians such as Robert Cox have argued, each historical structure helps generate the conflicts and tensions out of which new structures grow.[33]

However, even if attention is confined to the restricted level of analysis singled out by mainstream International Relations theorists, namely specific pressures exerted on states by virtue of their membership of a society of states (since one cannot write about everything at once), then there is much that the mainstream rationalist approaches tend to bracket, ignore, or marginalize. One of the distinguishing marks of a critical constructivist approach to world politics is that it has offered a framework for understanding precisely those dimensions of world politics that neorealism and neoliberal institutionalism conveniently bracket: the changing identities and interests of states, the role of moral norms, and the significance of culture and discourse in international politics. Whereas mainstream approaches to the study of international relations have traditionally considered these features to be epiphenomenal, critical constructivists maintain that they are central to understanding the socially constructed "reality" of world politics—including questions of material capabilities and interests. And whereas the mainstream rationalist approaches take the structure of the state system as fixed, and the interests and identities of states as given, a priori and exogenously, critical constructivists *historicize* both of these aspects of world society. In particular, they emphasise the role of social agency, moral entrepreneurs, and critical discourse in transforming both social structures and the identities and interests of social actors, which are understood as endogenous and socially constructed.[34] Of special interest to critical constructivists is not only the processes of regime formation but also what goes on before and after regime formation, such as how norms emerge, how states come to acquire their identities and interests, and how changes in shared understandings might serve to transform the meaning of the constitutive principle of state sovereignty. In exploring these questions, critical constructivists eschew the atomistic ontology of the rational actor model of mainstream International Relations theory and proceed instead on the basis of a relational ontology that can be best understood as a *social learning* model. In the context of this model, social actors (including governments acting on behalf of states in the international arena) are understood as reflexive actors that do not necessarily follow simple scripts or subscribe to formulaic calculations, nor do they simply respond or adapt to constraints provided by their external environment in accordance with fixed interests and preferences. Rather, through social interaction, such actors may also change their understanding of, and therefore their relationship to, their external environment (including other social actors) in ways that may transform their understanding of their own interests and identities. According to this social constructivist perspective, then, norms and rules do not merely coordinate and regulate the behavior of social actors with pre-given interests and identities; norms and rules can also *constitute or reconstitute* the interests and identities of social actors by defining or redefining the set of practices that make up social activity. These practices in turn define who is a legitimate actor in particular social contexts.[35]

In making these claims, critical constructivists work with a broader understanding of politics and hence are sensitive to a broader repertoire of political action by states and nonstate actors than rationalists. That is, state and nonstate actors may engage not only in coercive or strategic action but also in deliberation and persuasion ("communicative action"). While one can draw analytical distinctions among these different forms of political action, in practice, they are often enmeshed and are rarely a simple function of material capabilities. For example, hegemonic powers are not always immune to moral argument, while moral advocacy activists, including environmental NGOs, often engage in strategic action and sometimes even coercive action (e.g., ecosabotage). Moreover, to acknowledge this broader repertoire of political action is also to acknowledge that social actors may cooperate or otherwise conform to social norms or legal rules for a range of different reasons.

They may conform out of fear of punishment, or because they believe that following the norm/rule is in their own self-interest, or because they accept the norm/rule as legitimate.[36] Whereas self-interested behavior entails a continuous instrumental calculation of outcomes in accordance with pre-given, self-interested goals, the acceptance of norms as legitimate necessarily works at the intersubjective level. That is, a rule or norm is considered legitimate by an actor to the extent to which the actor "internalizes its content and reconceives his or her interests according to the rule."[37] Moreover, what might have started out as a purely instrumental calculation by social agents may end up transforming the self-understanding and identity of such agents.

It is therefore misguided to insist that constructivists can succeed in empirically demonstrating the sway of moral norms in world politics *only* when conforming with such norms would go against an actor's own interests, since this misunderstands the ways in which the actor's interests might have become realigned with broader intersubjective understandings, in which case there will not be a conflict of interest.[38] Moreover it is precisely when the self-interested behavior of actors shifts into alignment with collective norms of environmental protection that problems such as the tragedy of the commons are avoided.[39] Thus politics as a struggle for power and politics as social learning should not be understood as independent movements/dynamics, since "powering" and "puzzling" are often intertwined in the formation of public policies.[40] It is under circumstances such as these that rational choice theory struggles with behavior that defies the problem of collective action. One response, as I have noted, is to continually redraw the meaning of "self-interest" in broader and more iterative terms to encompass such dimensions as fear of loss of reputation, cognitive structures, ideas, and epistemic communities. However, such an approach makes the category of self-interest too protean to be of any value. That is, it is made to subsume all other potential categories of explanation to the point that it risks becoming tautological.[41] If the explanation of state's interests (understood as self-interests) is to illuminate, it must be framed in such a way as to be clearly differentiated from legitimacy explanations.[42]

Now critical constructivists do not argue that norms can bring about social transformation in a mechanical or direct causal way. Providing persuasive reasons for action is not the same as causing action.[43] Shared understandings resulting from moral persuasion, and a realignment of identities and interests have to be given effect by means of changes in material practices. Nonetheless, shifts in shared understandings provide the meaning and context in which shifts in material practices take place. As Daniel Philpott has persuasively argued, revolutions in sovereignty stem from prior revolutions in ideas about justice, political authority and rightful conduct.[44] Similarly the fact that a particular state happens to possess considerable material capabilities certainly creates the possibility of coercive practices by that state, and history is replete with examples of powerful states bending the will of weak states through explicit or implicit coercion. But this is not invariably the case, and whether material capabilities are likely to be deployed by states or other social actors in coercive ways can only be understood in the context of the histories and social relationships of self/other between *particular* social actors. Moreover such relationships cannot be simply deduced from social structures; they can only be understood in the context of what John Ruggie has called "narrative explanatory protocols."[45] This is an interpretivist account that looks back and makes sense of what happens rather than one that seeks to predict behavior according to law like generalizations.

Now once coercion, self-interest, *and* legitimacy are acknowledged as possible reasons for the behavior and interactions of social actors, including states, then in multilateral dealings, as Ian Hurd points out, there is no obvious theoretical or empirical reason why we should limit our understanding of state behavior and/or world politics to only one or two of these three possible modalities.[46] Nor is there any good reason for regarding coercion and self-interest as default explanations, as if culture and moral norms have only secondary importance, working as a kind of leftover explanation that may only be brought in when neorealist or neoliberal explanations cannot fully account for the behavior of particular states or the outcomes of particular multilateral negotiations. Such an approach places an unfair burden of proof on critical constructivists, who must demonstrate not only that moral norms matter but also that they matter to the exclusion of power and interest, in the sense that they must be shown to be untainted by these more "base" material interests in order to have any explanatory power. As Hurd argues, "We have no better reason to *assume* coercion than to *assume* legitimacy."[47] Coercion, self-interest, and moral argument are often deeply enmeshed in multilateral negotiations, and it is therefore likely to be an atypical situation when one of these modes of interaction can be found to exist in pure form. That is, a regime based on pure consent is as unlikely as a regime resting on pure coercion.

Indeed, any regime based on coercion alone is inefficient, unstable, and costly. Stable and enduring influence flows not from coercive power but rather from either *legitimate* power (based on free consent) or *hegemonic* power (based on a mixture of consent and coercion). Gramscians and neo-Gramscians argue that in the case of hegemonic power the justification for the order is universalistic rather than blatantly self-serving. In other words, the hegemonic class or state is able to persuade subaltern classes or less powerful states to see the world in terms favorable to its own ascendancy. Coercion is mostly latent and usually applied in a disciplinary way in marginal cases.[48] This helps explain why dominant states often engage in inconsistent behavior (i.e., swinging from unilateral and multilateral action) in different settings. As Bruce Cronin has explained, such states suffer "role strain" between their relative position as a *powerful* state (where they have the capabilities to act unilaterally and appease domestic social forces and interests to the exclusion of international concerns) and their role as a *hegemonic* state (where there are social expectations that they will confirm to generalized rules of conduct, which suit their longer term interests in maintaining a stable and legitimate international order).[49] On this understanding, hegemony is not simply a function of economic and military material capability; rather, it also turns on whether a state is able to shape the international order according to norms and rules that mostly suit its interests but are more or less accepted by others as universal. As G. John Ikenberry and Charles A. Kupchan have also argued, hegemonic power is exercised as a form of socialization that is manifest when foreign elites internalize the norms and values espoused by the hegemon and accept its vision of international order as their own.[50]

Likewise Joseph Nye has recently drawn a distinction between "hard power" (or command power resting on coercion or inducement) and "soft power" (cooptive power). Soft power involves the ability, through rule-based multilateralism (among other means), to shape what others want, by making one's culture and ideology attractive to others.[51] Like Cronin, Nye points to the longer term limitations of any sustained reliance on hard power, particularly in the current, intensely interdependent context of economic globalization and the rise of nonstate actors. However, Nye writes with mainly an American audience in mind, and his defense of soft power (and multilateralism) is mostly a pragmatic one. That is, it is sold as a smarter way of advancing American interests (compared to the muscular unilateralism of the second Bush administration).[52]

Nye's argument is important, but it neglects two crucial aspects of multilateralism that are of interest to critical constructivists. The first concerns the social and symbolic dimensions of *membership* in an international society, and the political obligations and relationships that attach to membership.[53] Even powerful states prefer recognition and approval from the international community (though they can, and sometimes do, assert themselves without that approval by enlisting hard power). Acting outside the web of norms and practices widely recognized as appropriate for a member of a political community brings censure and isolation. The fact that the second Bush administration attempted (albeit unsuccessfully) to argue the case for military intervention in Iraq in the UN Security Council and to the international community (rather than simply wage war on Iraq without at least attempting to persuade others of the rightness of its actions) is itself testimony to the importance of international recognition and the acquisition of legitimacy.[54] The second neglected aspect of multilateralism concerns the way that this need for recognition and legitimacy gives middle power and weaker states a particular kind of countervailing power vis-à-vis the hegemon, a power that is akin to the power of the slave over the master in Hegel's master–slave dialectic, the power to withhold recognition and support for actions that they consider cannot be generalized or justified. While this kind of countervailing power cannot withstand naked force, *in the long run* no powerful state can rely only on naked force alone if it wishes to remain a member and enjoy the benefits of the international society of states. From a critical constructivist perspective, "genuine" recognition for proposed norms or actions only arises when the assent of other states is achieved by means of reasoned dialogue—the unforced force of the better argument—or by exemplary action that engenders uncoerced approval and emulation by other states.

These insights into the nature of hegemonic power make it clear that states may cooperate for reasons other than, or in addition to, coercion.[55] It is precisely because of the difficulties of maintaining any regime or social order through brute force alone that hegemonic powers often find it necessary to defend their favored regime/social order in universalistic rhetoric, *by providing reasons that others may find acceptable*. But as soon as this move is made, hegemonic powers leave themselves open to being disciplined by their own rhetoric, even when it may be inconvenient or directly contrary to their more immediate interests. Yet for a powerful state or social actor to exempt itself from its own rhetoric is to risk loss of legitimacy (and possibly require resort to coercion to restore order).[56] Of course, legitimacy is always a matter of degree, and the more one moves from a social arrangement based mostly on self-serving legitimations by hegemonic powers to a genuinely legitimate social order (whose organizing principles are reflectively acceptable to all), then the more we can expect the social arrangement to set real limits on the conduct of the powerful as well as the powerless. As David Beetham has put it, "it is power itself that morally stands in need of legitimation, though not every form of power requires it in practice, and by no means all achieve it."[57]

Beetham's sociological observation also points to the epistemological congruence between the explanatory purpose of the social scientist and the emancipatory purpose of nonhegemonic or subaltern groups wishing to understand social structures in order to transform them.[58] This also encapsulates the emancipatory aim and method of immanent critique adopted in this inquiry. However, the Catch-22 is that the task of exposing contradictions and self-serving arguments from the perspective of those who benefit least from a particular social order is made easier to the degree in which the communicative context is unconstrained, which it typically is not.

Now a predictable objection might be raised at this point that the communicative context for multilateral negotiations is typically distorted in significant ways (due to highly unequal bargaining power and the repeated flexing of "muscle" by powerful states, imperfect knowledge, incomplete time for deliberation, problem complexity, cultural difference, and in some cases cultural incommensurability, and lack of social inclusiveness in terms of nonstate actors) such that we can expect communicative action to play only a relatively minor role compared to coercion and strategic bargaining. Critics might go on to point out that the ideal of communicative justice assumes the very absence of those things that still dominate international relations: power and self-interest, not legitimacy. Indeed, this view is shared by many deliberative democrats and neo-Habermasians who consider that Jürgen Habermas's communicative ethics is fundamentally at odds with the practical discourses and procedures of regime rule-formation among sovereign states in international relations or that it only has application to domestic society.[59]

However, such an argument misconstrues the role of unconstrained communication as a counterfactual ideal and critical vantage point from which to judge social interaction. A "counterfactual ideal" is a claim as to what could or might happen *if* certain communicative conditions were fulfilled. In this respect it can illuminate the real world by providing a critical vantage point from which to evaluate the degree of distortion of particular communicative contexts (accepting that all real world communication at best can only ever be an asymptotic approximation to the ideal). Far from removing power from the equation, Habermas's counterfactual ideal enables us to observe the many ways in which the *presence* of power can distort communication. In any event, international society is not *so* removed from this counterfactual such as to render the ideal irrelevant. While traditional realists are right to point to the presence and exertion of material power in international politics, they are wrong in assuming the total absence of any common life-world in the international realm. As Thomas Risse has argued, even an anarchic world order is based on a "thin" common life-world based on shared meanings, and histories among many nations, most notably, World War II and shared ecological problems.[60] Moreover a focus on legitimacy explains why it is that most states observe international treaties most of the time in the absence of a world police force. Neo-Gramscian critical theory also explains why this often includes hegemons, even in circumstances where conformity may go against the hegemon's narrowly conceived interests.[61]

Indeed, in his more recent work Habermas has emphasized the inevitable and enduring tension between the idealized presuppositions embedded in communication and the practical exigencies of real world communication. This is part of what Habermas means when he refers to the tension between "validity" and "facticity," between the ideal and the actual. While he developed this understanding in the context of the tensions between the regulative ideals of democratic law making and the actual production of positive law at the level of the domestic state, it can also shed light on multilateral law making of states, particularly in view of the increasing role played by international civil society.[62] Habermas himself has not explicitly extended his discourse theory of law to the international domain in this way, although he has recently argued that transnational public spheres have increasingly become sites for the discussion of global issues.[63]

Beetham's observation also encapsulates the Habermasian argument that all (rational) communication is at least implicitly *oriented* toward reaching mutual understanding by means of persuasion rather than coercion or bribery, even if such understanding is not *actually* reached. Even in highly distorted communicative settings, parties can still feel obliged to give reasons for their preferred positions if they are to persuade others of the acceptability of their arguments in order to reach an agreement or simply to be recognized as legitimate actors, even if no agreement can be reached. And, as I have already noted above, the critical testing of claims—which can, on occasion, include public shaming—is also one of the few weapons of the weak (others might include withdrawal from social cooperation, e.g., labor strikes and consumer boycotts, or even hunger strikes). There are very few states in the world that are completely impervious to such public exposure and critical dialogue. Even hegemonic states bent on exerting their will still find it necessary to give reasons for their international actions. Success in such argumentation is a function, among other things of the degree of trust, truthfulness, and respect among the parties, the character of their shared history, and whether parties have the capacity to perform the promises they undertake. While the existence of one very powerful state that chooses to act unilaterally can serve to distort the ideal of responsible membership in international society based on common understandings about rightful conduct, this need not be the inevitable response of a superpower. The second Bush administration's repudiation of the Kyoto Protocol (as well as the Convention against Landmines and the International Criminal Court) should be understood as the predilections of a particular administration rather than the necessary response of a superpower. The United States has not always acted unilaterally, and it is perfectly capable of exercising its power in more responsible ways—it is just that *this* administration has chosen a different path based on its own perceptions of America's future energy needs (despite the fact that public opinion in America favors ratification of the Kyoto Protocol).[64] Ironically, as Nye points out, unilateralism—even by a superpower—is likely to be ultimately self-defeating in the longer run in a world of complex interdependence. The Kyoto Protocol is likely to be ratified despite the lack of US support, and the United States may find, further down the track, that it has missed an opportunity to exploit the longer term ecological and economic benefits of pursuing an energy strategy based on ecoefficiency and ecological modernization.

2.3.1 Not One but Many "Cultures of Anarchy"

One of the limitations of neorealists' and neoliberal institutionalists' explanations of international politics is that each assumes that there is only *one significant* culture of international anarchy (i.e., Hobbesian and Lockean respectively) in multilateral negotiations. However, if we think of the international community as layered, as made up of many different communities, cultures, and associated modes of relating to the other, then we must necessarily acknowledge the complex multicultural rather than unicultural character of international relations. Alexander Wendt's analysis of the different cultures of anarchy in the international community is helpful in this regard as it explores the sociological phenomena of relating to, arguing with, and responding to others in the context of history. Just as different social structures can produce different social roles and identities, and different modes of relating, so too can different cultures of anarchy produce different state roles and relationships. For example, Wendt shows how states may relate to other states as enemy, rival, or friend, and these roles correspond to three different cultures of international politics—Hobbesian, Lockean, and Kantian (these are identified by Wendt as "salient" logics and are therefore need not be taken as exhaustive). Moreover these different cultures of anarchy explain why states, when they inhabit certain roles, conform to certain behaviors or norms. That is, when they relate to other states as enemies they are only likely to "cooperate" with others when implicitly or explicitly coerced, when they relate as rivals they tend to comply mostly out of self-interest, and when they relate as friends they comply principally because of shared, internalized understandings. We should also expect that moral/ethical reasoning will, potentially at least, play a bigger role than instrumental reasoning in political communication among friends, given the depth of shared understandings, but this is not a point that is explored by Wendt. It is also quite possible, however, that instrumental reasoning would continue to dominate against a background of shared moral/ethical understandings (whereby what is agreed/shared operates in the background and the disagreements occupy the foreground of negotiations).

Wendt makes it clear that the existence of a Kantian culture of relating among sovereign states need not necessarily imply that there are not important differences and disagreements among states; rather, it simply means that states mostly relate to each other as friends rather than rivals or enemies. Here "friendship" is understood as a "role structure" whereby disputes are settled without war or threat of war *and* mutual aid is provided to members in the face of external threat.[65] This relationship of friendship is said to be more enduring than the relationship among members of purely strategic alliances, which are more contingent, precarious, and liable to fall apart with shifts in the balance of power.[66] Friendship is based on a shared knowledge and history of the other's peaceful intentions. In such circumstances cooperation cannot be reduced to material self-interest but can only be understood in terms of the mutual *internalization* of shared norms. That is, the conception and welfare of the self is taken to include others in the community.[67] However, this identification with the other is rarely total since actors, including states (through their negotiating agents), typically have multiple identities.[68] We can therefore expect contestation and some resistance to arise among members in efforts to reach shared understandings, including debates about free riding and burden sharing in any negotiations over common environmental problems. If communities repeatedly fail to resolve such differences, we might expect the Kantian culture to wane over time.

Now Wendt confines his elaboration of the "culture of friendship" to collective security communities, and suggests that the trust and generalized reciprocity that they entail need not necessarily spill over into other issue areas, although the observance of nonviolence and mutual aid sets limits on how other issues are likely to be dealt with.[69] Yet in principle there seems to be no reason not to expect this culture of shared understandings to frame or at least have some influence on other issue areas, and we would expect members of a "Kantian security team" to find it easier to reach agreement about other common problems, such as environmental problems, with friendly states than with other states that may be rivals, enemies, or just strangers. Moreover we would expect the extent to which environmental cooperation might occur to be not only a function of mutual respect but also a function of the more open character of the discursive processes within the community of friends.

Against this background, the North Atlantic Treaty Organization (NATO) is an interesting case. The stark differences between the United States and Britain, on the one hand, and continental Europe, on the other, over the question of intervention in Iraq have placed further strains on the security community of NATO, whose rationale had already become uncertain following the collapse of the iron curtain. Yet these differences are symptomatic of deeper differences across the Atlantic. The US position as the unrivaled super power in the post–cold war period, combined with the increasing economic and political integration of Europe, has led to a growing gap between the United States and the EU over their ideas of world order and their respective commitment to multilateralism, including environmental multilateralism. In this context the EU operates as a subcommunity not only within NATO but also in the field of environmental diplomacy.

Like Wendt, Andrew Linklater has also emphasized the multiple and overlapping character of different frameworks of relating among states. Building on the postnationalist explorations of E. H. Carr, Linklater has identified three frameworks of action among states that go beyond realism, which he calls pluralist, solidarist, and post-Westphalian.[70] Pluralist communities operate under a framework of toleration and mutual respect for difference among radically different states that desire to preserve their autonomy, whereas solidarist communities refer to a framework of action between states that share certain moral and ethical principles (e.g., commitment to human rights). These two categories are not identitical with Wendt's Lockean and Kantean cultures, but they nonetheless resonate strongly. However, states belonging to a community that conforms to Linklater's third, post-Westphalian category have taken a step beyond the Kantian community by forgoing some of their prerogative powers as states in order to deepen cooperation and collective problem solving, thereby breaking the traditional assumption of congruence between sovereignty, nationality, citizenship and territoriality.[71] Linklater is careful to point out that states (through their negotiators) have the hermeneutical skills to recognize the heterogeneity of international society and would not assume that the post-Westphalian arrangement is inherently superior to pluralistic or solidarist arrangements for all states since much depends on the degree of cultural compatibility among states.[72] Different arrangements—pluralist, solidarist, or post-Westphalian—may be appropriate in different contexts and may be best approached by dialogue that avoids imperialist pretensions. However, post-Westphalian arrangements emerge as real possibilities when certain normative preconditions exist, which Linklater identifies as a commitment to not only deliberative dialogue (understood to include respect for difference), but the social and ecological preconditions for such dialogue, which include constitutionalism and the rule of law.[73] These preconditions, Linklater argues, represent the positive features of modernity that enable a unit-driven, peaceful transformation of the international order.

Now Wendt's and Linklater's understanding of the diversity of "cultures of international anarchy," like Habermas' theory of communicative action, makes no predictions about international politics, although Wendt does suggest that at this juncture, the international behavior of states is mostly Lockean (rather than Hobbesian) but with increasing Kantian (or we might add post-Westphalian) dimensions.[74] This certainly helps explain why neoliberal institutionalism has become the dominant framework for analyzing environmental regimes to date (since it does resonate with the actual behavior of many states), while also historicizing this dominance and pointing to new directions for political research and political action. Moreover Wendt's and Linklater's frameworks can be usefully applied to different cultures of anarchy in ways that build on rather than dismiss the insights of neorealists and neoliberals. We would also expect to find different states belonging to many different types of community. Even in Hobbesian and Lockean cultures, historical relationships and shared understandings provide the key to understanding how material capabilities and interests are interpreted.

It should be emphasized that none of the arguments above deny the force of the eco-realist analysis *in those domains where a Hobbesian culture prevails*. But a Hobbesian culture does not always prevail and it is not always dominant. Here I would agree with Wendt's reading that the international behavior of states is mostly Lockean but with increasing Kantian (or post-Westphalian) dimensions.[75] Moreover, for neorealists or neoliberals to deny or dismiss the possibility of something like a Kantian culture of responsibility, or a post-Westphalian framework of action taking a stronger hold on international society, is to make the politically conservative move of sanctifying the Hobbesian or Lockean culture as somehow natural and immutable. My critical ecology perspective, in contrast, would search for ways to transform extant cultures toward a Kantian or post-Westphalian culture of relating in those circumstances where the trajectory of historical relationships and shared understandings make this a genuine and desirable possibility. The European Union is probably the closest empirical approximation of a greenish Kantian culture, with intimations of a post-Westphalian culture, and this is partly explicable by the close geographical proximity of states and a shared history particularly in relation to security and ecological problems. From relatively modest beginnings in the European Iron and Steel Community of 1955, the EU has developed into a transnational polity, with its own regional Parliament, Court of Human Rights, common trade and foreign policy, and common currency. Moreover its common security and ecological problems have generated new ecological discourses about security, sustainable development (and ecological modernization) that have extended to include common environmental rights and environmental justice norms. Of particular interest is that many of the new understandings generated by these discourses have *already found expression* in multilateral agreements within the EU that have served to facilitate ecological citizenship within the region.[76] Moreover the EU has acquired a distinctly green identity and has played a leadership role in international negotiations over major global ecological problems such as climate change. As I show in chapter 7, states belonging to such a post-Westphalian community may be described as "transnational states" rather than nation-states.

The new and interlinked ecological discourses are both strategic *and* moral discourses, reflecting new issues, agendas, and values about the ultimate meaning and purpose of individual and social life. Some of the counterhegemonic green discourses provide nothing less than a civilisational critique of modern life that carries intimations of a new ecologically just world order where ecological (and social) risks are drastically minimized and localized, and where there is no undue disparity in the "ecological footprint"[77] or "ecological shadow"[78] left by different classes, states, and regions. These intimations of an ecologically just world order carry with them a set of radical reconceptualizations of what amounts to legitimate use and illegitimate abuse of property and territory, and human and nonhuman nature, by human agents and social collectivities (including states). Such discourses challenge the traditional rights of private property holders as well as the territorial rights of states to conduct activities that compromise the ecological integrity of the nonhuman world both within and beyond their territories.

2.3.2 Toward Structural Transformation?

Of particular interest to critical political ecology is whether these new developments in environmental multilateralism carry the potential to transform the international order, including the organizing principle of sovereignty. Whereas the realist tradition exposes (and is largely resigned to) the deep ecological irrationalities of the principle of exclusive territorial rule, and whereas the neoliberal institutionalist approach merely helps us see their limitations and find ways of ameliorating them, neither of the mainstream approaches suggest how these irrationalities or limitations might be addressed in a fundamental way. Both seem to be blind to emergent developments that enable new conceptualizations of the sovereign territorial rights of states.

Reflecting on these deep changes to the international order, John Ruggie has argued that what we are witnessing is a certain "unbundling of territoriality" that has challenged the current system of governance by discrete enclaves.[79] That is, the Westphalian system of sovereign states is generally understood to have established a particular *system of rule* that "has differentiated its subject collectively into territorially defined, fixed, and mutually exclusive enclaves of legitimate dominion."[80] The connection here is between the conception of absolute and exclusive private property based on Roman law and the conception of absolute and exclusive sovereignty.[81] Our concern here is that this international structure of rule has managed to reproduce itself in the face of what Ruggie has called "the paradox of individuation." This paradox arises from the fact that a system of rule that is territorially defined, fixed, and mutually exclusive has no ready means of managing (1) territory or spaces falling outside the territorial jurisdiction of states (e.g., oceans, waterways, and the atmosphere) or (2) problems of common concern that are irreducibly transterritorial in nature (e.g., global warming). Realists seem resigned to this state of affairs, since ecological destruction is a "natural," in the sense of inevitable, product of the system of sovereign states. However, from the perspective of critical political ecology, these two examples are paradigmatic of most ecological problems, and they suggest that the ecological crisis has the *potential* to transform the rationale and structure of exclusive territorial rule, and the identities and interests of states (but without necessarily dispensing with the principle of sovereignty, as I show in chapter 8).

According to Ruggie's analysis, *multilateralism* has resolved "the paradox of individuation." That is, states succeed in reproducing themselves by virtue of the development of multilateral norms and institutions, which may be understood as providing an evolving supplementary structure of rule that compensates for the limitations of an exclusively territorial structure of rule. Indeed, Ruggie regards the modern international polity—the complex layers of multilateral norms and institutions that govern international society—as "an institutional *negation* of exclusive territoriality."[82] To what extent a supplementary structure of rule in the environmental domain will eventually serve to transform the (territorial) structure of rule on which it rests is an open question. However, anticipating, imagining, and critically interrogating, the possibility of such a transformation should also be one of the tasks of critical political ecologists concerned with hunting down "emancipatory openings."

While neoliberal institutionalists recognize the problems generated by complex interdependence, their approach challenges neither the state's exclusive territorial rights nor the idea of states as individuated units that are merely externally related. They acknowledge that the "self" or individuated unit (understood as the individual, the community, the state or the bioregion) can no longer "rule" or exercise autonomy effectively without some accommodation of interdependence and a broader set of transboundary/common concerns and responsibilities. However, this acknowledgment only becomes an ontological and epistemological breakthrough when the basic units of the system are understood as internally rather than externally related. Once this breakthrough is made, we move away from a classical liberal understanding of freedom or self-rule toward a model that is closer to the civic republican understanding of civic virtue and responsibility for the common good. According to the republican understanding, freedom is something that is *constituted* by mutually negotiated and mutually recognized norms, or common rules, whereas for the classical liberal the social contract merely sets up a limited legal framework that authorizes a justified but limited *interference* with preexisting "natural" rights. A critical political ecological understanding of autonomy or self-rule (whether of individuals, community associations, or abstract communities such as nations) would likewise be one that is constituted by shared norms, rules, and identities, at least at the level of ecological awareness. The daring idea that common norms and rules might one day constitute nation-states as "local agents of the common good,"[83] custodians of the biosphere, public trustees, or planetary stewards may appear fanciful to neorealists and neoliberals, but the intimations of such a posture are already present in many multilateral and regional environmental declarations, strategies and treaties (as I explore in chapters 7 and 8). These developments need not entail any relinquishment of sovereignty, but they may entail its reconceptualization in less exclusive terms in the relation between peoples and territories. This reconceptualization would also entail a broadening of the identities—indeed the very *raison d'être*—of states so that they are understood in the primary rather than subsidiary sense as custodians of the biotic community and not as exclusive property holders. Such claims, of course, can only have the character of a promissory notes at this stage. However, in chapter 8, I track recent developments in international environmental law and policy, show how new norms of ecologically legitimate state conduct might be coaxed out of the existing environmental multilateral order, and suggest how they might be defended.

In this chapter I have pointed to the potential of the nation-state to take on the role of ecological trustee, thereby shedding its traditional role of environmental exploiter and harmonizing its role of both developer and environmental protector. Unlike mainstream International Relations theories (i.e., neorealism and neoliberal institutionalism), the critical constructivist perspective that I have built on is open to the possibility of states taking on such a role and promoting this development. In contrast, mainstream International Relations theory has not allowed ecological problems to encroach upon their assumptions and habits of thought, least of all their understanding of multilateralism and the principle of state sovereignty. Against mainstream theory, I show in chapter 7 how nation-states have already become transnational-states that serve a broader community than the nation, while in chapter 8, I point to the ways in which new developments in international environmental law and policy have given rise to green evolutions in sovereignty.

However, any deeper greening of states presupposes the alleviation of *other* systemic, anti-ecological pressures on states. I refer here not to the systemic pressures that arise from a Hobbesian or Lockean anarchic system of states per se but rather to the systemic pressures arising from the development of global capitalism, which are increasingly being expressed *through* the state system in *economic* multilateral arrangements covering trade, finance, debt relief, technology, and development. The biggest challenge to the development of green states comes not from pressure generated by the state system but rather from the competitive pressures of global capitalism. The potential for more innovative initiatives in *environmental* multilateralism is therefore likely to be limited until such time as the ecological contradictions generated by dominant *economic* multilateral arrangements are resolved by more reflexive, and hence more ecologically sensitive, modernization.

3

The State and Global Capitalism

3.1 The Decline of the State?

A central issue in the contemporary debate about economic globalization and the state is whether the state is in some form of possibly terminal decline, or whether it is merely being transformed or reconstituted.[1] While nation-states still constitute the principal form of political rule throughout the world, the idea that they occupy the "commanding heights" of governance, effectively and legitimately regulating and insulating social and economic life within their borders, has fallen by the wayside. It has been widely observed that territorial borders have become less significant and that states are increasingly subject to global and regional forces beyond their control.[2] In particular, it has been claimed that the welfare functions of the state are being made residual to the primary tasks of the new "competition state" operating under the dictates of what Stephen Gill has called "disciplinary neoliberalism."[3] According to this critical perspective, the state's "traditional" welfare services and its "emergent" environmental services are increasingly seen to be brakes on economic growth. Although the state lives on, these developments do not augur well for the emergence of green democratic states that are both willing and able to uphold public values, and to act as vehicles for social and environmental justice.

In this chapter I show that the changing form of global capitalism creates a situation that is more complicated but not quite as bleak as my preliminary analysis in chapter 1 suggests, at least for most developed states. By highlighting the variability in the adaptive strategies of differently situated states to new competitive pressures, I depart from the bleak orthodoxy in critical thought and offer a cautiously optimistic assessment of the evolving relationship between states, global capitalism and civil society.

As early as the 1970s, neo-Marxist theorists drew attention to the "fiscal crisis" of the welfare state stemming from the state's contradictory imperatives to facilitate capital accumulation, on the one hand, and to iron out the harmful social and ecological consequences of capital accumulation by providing an expanding menu of protective welfare (and environmental) services, on the other hand.[4] Now, in the new millennium, the growing intensity of economic regionalization and globalization is making it increasingly difficult for governments to solve a range of social and ecological problems within their territory and beyond.

In this chapter I track these developments by working through three different critical frames and phases of understanding of the changing role and function of the liberal capitalist state. I begin with the neo-Marxist analyses of the contradictory functions of the welfare state in the 1970s, move to the specifically critical ecological analyses of the state in the 1980s and 1990s, and then to the more globally oriented, critical analyses of the state in the 1990s and beyond. After addressing concerns about declining state autonomy and the problem of the competitive "race to the bottom" in environmental standards, in the last two sections of this chapter I critically explore the new discourse and practice of ecological modernization as an adaptive strategy of aspiring green states responding to competitive pressures.

My concern is to restore the dual focus of critical theory on the state as a site of not only environmental exploitation but also environmental protection. Consistent with the aims of critical political ecology, I explore what it might take to dampen the exclusionary dimensions and promote the inclusionary possibilities offered by ecological modernization.

3.2 Eco-Marxism, the Welfare State, and Legitimation Crisis

Critiques of the capitalist welfare state in the 1970s focused on the state's fundamentally contradictory tasks.[5] On the one hand, the state was seen as having an institutional interest in safeguarding the interests of capital. This was not because of any ruling class conspiracy but rather because of its functional dependence on the flow of revenue (principally taxation) that private capital accumulation provides.[6] In this respect the state was understood as defending and upholding the interests of a capitalist *society*, including workers, investors, and consumers, rather than merely the interests of the capitalist *class* standing alone. Capitalist states typically do this by providing the necessary legal and social infrastructure for businesses to flourish, as well all those facilities and services that contribute to the growth of capitalist society.

On the other hand, the state was posited as having to respond to public pressure to redress the negative social and ecological "side effects" generated by private capital accumulation (including the commercial activities of state-owned instrumentalities). If the liberal capitalist state was to be seen as representing all of society, then it had to be responsive not only to the demands of capital but also to the demands of all those who are exploited or otherwise harmed by capitalism's tendency to privatize gains and socialize costs. Yet the state's capacity to secure its legitimacy by alleviating these problems via its welfare and protective services is typically dependent on its *also* performing successfully the function of maintaining private capital accumulation. So while the formal rules of liberal representative democracy enable this legitimation function to be discharged (to some extent), the boundaries of successful policies are invariably set by the buoyancy of the economy. The upshot was that any concerted attempt to regulate private investment and business activities to the point where negative ecological externalities are eliminated or made negligible was believed to bring about a set of multiple crises, for instance, inflation, capital strike or flight, and labor unrest. The core claim in my highly schematic account of this body of state theory was that these contradictions—to provide for the interests of private capital *and* to dampen social unrest by ironing out the negative social externalities of capitalist accumulation—cannot all be resolved simply by pursuing more efficient or more effective economic management and administration. Rather, these tensions can only be "politically managed" because few governments are prepared to risk serious economic dislocation or any cessation or major curbing of economic growth in the name of environmental protection: to do so would merely hasten their political demise.

The systems-theoretic approach employed in the early work of leading theorists such as James O'Connor, Claus Offe, and Jürgen Habermas focused on the *functional interdependencies* between the capitalist state and the capitalist economy. That is, the basic unit of study is not the state as an independent or historical entity but rather the *state capitalist system*.[7] (This may be contrasted with non-Marxist theories of the state that approach the state as an "autonomous" organization or source of power "controlling, or attempting to control, territories and people."[8]) These functional interdependencies are examined with a view to ascertaining the *policy limits* and *policy failures* of the capitalist state.[9] Since policies are understood in terms of the functions they serve vis-à-vis the broader system of which they are part, policy failure (whether in the domains of social or environmental policy) may be understood as one particular instance of system failure. For example, environmental policy failure can be traced to structural contradictions in the capitalist economy, which are analyzed as giving rise to crisis tendencies.[10] It is precisely this functional dependence on the processes of private capital accumulation that makes the welfare state a "capitalist state" and sets limits to the scope and substance of state policy making.[11] The welfare state is understood to be in *crisis* because these contradictory requirements of accumulation and legitimation can never be resolved within the system's own boundaries.[12]

Although these early theories of the state did not make a special feature of environmental problems in their analyses, such problems could be readily lumped into the general category of negative side effects (alongside poverty and unemployment) generated by the processes of capital accumulation. Note, however, that the systems theory framework rested on the basic assumption that economic growth and environmental protection were zero-sum games. That is, more capitalist economic growth meant more environmental degradation while more environmental protection was assumed to mean less economic growth. This was consistent with the dominant understanding of the relationship between economic growth and environmental protection in the 1970s in the aftermath of the "limits-to-growth" debate.[13] Against the background of the oil crisis and highly publicized neo-Malthusian's predictions of future ecological catastrophe, calls by environmentalists for zero-growth or for a steady-state economy appeared as environmentally rational responses to a set of alarming projections. Of course, such calls appeared deeply irrational and politically suicidal from the perspective of state elites (politicians and bureaucrats) bearing political and administrative responsibility for economic management and employment.

However, during the 1980s and after, the assumption concerning the zero-sum relationship between economic growth and environmental protection was challenged. It was now widely acknowledged that there was some room for the development of virtuous synergies between economic growth and environmental protection. Exactly how much room, of course, remains a matter of debate between latter-day limits to growth advocates and ecological modernizers. Nonetheless, at least *to the extent* to which such synergies could be exploited, other things being equal, the basic accumulation/legitimation contradictions identified by critical theorists would be eased.[14]

Yet, during the 1980s and 1990s, the basic analysis of the contradictions of the capitalist welfare state was firmed in the face of the sustainable development debate.[15] James O'Connor's ecosocialist theory of the "second contradiction of capitalism" continued the long-standing Marxist understanding of the dynamics of capitalism as essentially contradictory and therefore containing the potential seeds of its own destruction or transformation. This potential, O'Connor argued, is manifested in the emergence of new social movements (notably environmental movements) that seek to challenge the destructive tendencies of capitalism. However, the environmental demands on the state are typically deflected, ignored, or dampened down by the capitalist state whenever they threaten the imperative of capital accumulation.[16]

For O'Connor, the basic dynamic of capitalism is one that continually undermines the social and ecological conditions for its own ongoing existence, a process that he articulated in terms of two fundamental contradictions. The first contradiction of capitalism refers to the contradiction between social production and private appropriation (the demand side), whereas the second contradiction refers to "the conditions of production" (the supply side), which O'Connor takes to be nature, labor, and infrastructure. Given the expansionary dynamic of capitalism and the limited supply/character of the conditions of production, he reasons that we can expect the costs of production to increase over time. This is exacerbated by the demands of labor, environment, and welfare movements to improve working conditions, protect the environment, and improve social infrastructure. For O'Connor, the so-called limits to growth do not appear as physical shortages but rather as higher costs. The contradiction arises from capital's standard response to the profit squeeze: to externalize costs. Yet such a response only serves to further reduce or undermine the profitability of the conditions of production and thereby raise the average costs of production.[17] The second contradiction of capitalism is held up as providing the most likely second road to socialism, although this time around it would be *ecological* socialism.

O'Connor argued that the combined effect of the first and second contradictions of capitalism is falling demand (from unemployment) and rising costs (from the limited supply of the conditions of production)—a problem that capital seeks to avoid by, for example, investing in nonproductive financial markets, which increases the vulnerability of economies. In all, he argued that there are few incentives for capital to be ecologically responsible in boom times, and even less so in recession or depression. The grow-or-die rationality of capitalism makes it crisis-ridden.

Against this background, the role of the state—and of policy makers generally—is to "rationalize" the conditions of production by improving the productivity of labor, protecting and regulating access to nature, or producing capitalist infrastructure.[18] The more the state undertakes such rationalization, however, the more the costs of production increase and the conditions of production become *socialized* in the form of more coherent state environmental planning and the technology-led restructuring of industry. But the outcome is uncertain, since the conflictual character of the pluralist policy process in capitalist states (which also favors the powerful and tends toward messy compromises) is such that no *systemic* resolution is likely. Resolving the contradictions requires integrated and coherent social and ecological planning on a scale that is beyond the motivation and capacity of the capitalist state. O'Connor argued that not only is the policy process too conflict ridden to achieve sufficient political unity, the bureaucratic state is also too fragmented and democratically insensitive to carry out such a momentous task.[19] Carrying forward the functionalist analysis of his early work, such a full-scale resolution of the ecological crisis is understood to lie *beyond the policy limits* of the capitalist state. While O'Connor acknowledged that capitalism has the potential to become more efficient in terms of material-energy use and waste production, he suggested that this is likely to be overshadowed by capitalism's attempt to "remake nature" to ensure sustainable profitability (e.g., by enlisting new technologies, such as those used to produce genetically modified crops). In short, capitalism is not, and cannot become, ecologically sustainable because capitalism can only expand or contract—"it cannot stand still."[20]

Yet O'Connor did not rule out the possibility of a progressive transformation of both the capitalist state and capitalist society into something different, and greener. To the extent to which such a mutual transformation is likely, it is expected to come from a democratic alliance of new social movements (including labor *and* environmental movements), mobilizing both the state and civil society to produce a new ecosocialist state and society. Far from withering away, O'Connor's ecosocialist state would play a crucial and considerable role in orchestrating an ecologically sustainable society. Exactly what form and rationale this state would take, and how it would manage what is admitted to be a challenging task, is not addressed in any detail.

John Dryzek's analysis of the ecological potential of the liberal capitalist state is broadly similar to that of O'Connor's, although Dryzek was even more sceptical than O'Connor about the prospects of the state ever acting as an agent of ecological emancipation. For example, Dryzek argued that capitalism, liberal democracy, and the administrative state work together to *compound* ecological problems.[21] In particular, the capitalist economy "imprisons" both liberal democracy and the administrative state, restricting its margins of successful policymaking and "punishing" those policy makers when they seek to step outside these margins. And within these narrow margins, the respective problem-solving rationalities of liberal democratic policy making and state administration tend toward problem displacement rather than problem resolution.[22] More recently Drzyek has pressed further his analysis of state imperatives, using it to explain the degree to which the inclusion of social movements in state processes of policy making is likely to be effective in terms of policy outcomes and democratization. In effect he argues that the democratization of the state via the inclusion of civil society actors is only really possible to the extent that the interests and claims of civil society actors accord with the state's functional imperatives. If the interests and claims of civil society actors do not accord in this way, then one can expect inclusion in the state to amount to cooptation, leading to the depletion of the unrestricted interplay of critical opposition in the public sphere.[23]

O'Connor's and Dryzek's basic analysis has been echoed by other political ecologists who have worked the familiar "contradictions" argument of critical theory at many more levels. For example, Colin Hay has argued that there is a fundamental mismatch between the level at which ecological contradictions are generated (based on the growth imperative of globalized capital accumulation) and the level at which political responsibility and crisis management is allocated (the liberal democratic nation-state).[24] And it is precisely because states, acting alone, are seen as incapable of resolving the crisis that they must develop instead "a complex repertoire of environmental *responsibility-displacement strategies*."[25] In effect, they seek ways of securing the state's legitimacy without actually resolving the underlying problems. Echoing Ulrich Beck's critique of contemporary risk management practices and Martin Jänicke's case of "state failure," Hay seeks to expose the limitations of "symptom alleviation, gesturing and responsibility displacement downwards (to individuals), upwards (to supranational institutions) or side-ways (other states)."[26] Whereas O'Connor and Dryzek hold out some hope for the redeeming qualities of the critical public sphere and for the possibility of new social movements acting back upon the state (a point I explore in more detail in chapter 6), Hay's ecological critique of the state is more devastating and his conclusions more pessimistic.

The foregoing theories of the capitalist welfare state have made a major contribution to the understanding of states as not isolated governance structures but rather as social structures that can only be understood *in relation to* society and the economy. However, such highly functionalist analysis can tend toward an overly deterministic understanding of state/economy relations. That is, functionalist claims can be difficult to prove or disprove, since any policy output can be explained as promoting the accumulation imperative or the requirements of legitimation. Short of catastrophic collapse, there are no other explanatory options available in this systems analytic framework. Partly in response to criticisms, Habermas and Offe have separately sought to defend their earlier theories as merely analytical or heuristic rather than empirical.[27] This is certainly a better way of understanding these functionalist theoretical claims—whose plausibility turns on the soundness of the author's analytical understanding of the state's functions. Yet, as Clyde Barrow points out, it is not possible to "explain the genesis of institutions and policies on the basis of imputed functions alone."[28] Nor can such theories be used to *predict* change or explain historical agency and the social meanings attached to such agency since "the axiomatic method employed by systems analysts continues to obstruct the transition from system to agency and from logic to history."[29] This problem also surfaces to some extent in O'Connor's more recent ecosocialist analysis of the state, which depicts economic forces as having an immutable objective logic, whereas political forces are contingent and indeterminate, the product of unpredictable discursive and power struggles.[30] As Stuart Rosewarne has pointed out, O'Connor attempts to force the different logics of the economic (objective, functional) and the political (intersubjective, contingent) into the one frame in a way that undercuts his long-standing effort to include social agency in social formations, including the "social construction of capitalism."[31] To frame these social struggles as merely reactions to "objective" developments is "to rob the social forces of any initiative."[32]

In a recent review of the general literature on theories of the state in relation to the sustainability challenge, Frederick Buttel has noted a steady movement away from what he calls deductive, nomothetic theories of the state, that is, theories that posit "some fundamental logic of state action generic to all states (or subcategories of them)."[33] Instead, Buttel calls for more attention to be directed to the political rather than economic logic of social orders, including discursive hegemony and the fragile nature of discourse coalitions. This accords with the critical constructivist perspective defended in this inquiry, which understands the material world as always already framed, contested, and mediated by the social world. This is not to replace historical materialism with historical idealism. Rather, it is simply to argue that there is nothing objective or deterministic about the development of the productive forces precisely because the meaning of these developments are unavoidably evaluative, historically contingent, and filtered through different social frames and social standpoints.

Now both O'Connor and Dryzek have also separately addressed sustainable development in terms of discourse theory rather than historical materialism.[34] Yet they still adhere to a functionalist understanding of the state. To be sure, Dryzek, for example, merely seeks to offer generalizations rather than "iron laws," but he still refers to the state's functions in terms of "imperatives," *as if* they were beyond the control of social agents, which sits uneasily with the ambivalent hope he invests in the green transformatory potential of new social movements and new discursive democratic designs.

In accepting the general heuristic value of these critical theories in relation to capitalist welfare states and the possibilities of political transformation, I consider that it is necessary to shift to a different level of abstraction. To this end, in the last two sections of this chapter I consider the role of *political and discursive struggles* over the contested meanings, purposes, and functions of social institutions. If the state as an institution is understood merely in terms of its objective functions, and policies are understood merely as strategic responses to systemic effects, then we have no social context that can explain why and how some policies are selected over others. The state thus "remains a black box of systemic conversion processes and abstract selective mechanisms."[35] The strategic intelligence of the state resides somewhere inside this black box, but is it a shadowy class of functionaries or state elites? High-level abstractions may direct attention to the broad parameters of policy making in liberal capitalist states, but they also obscure those things that are likely to shed light on social transformation, short of system collapse. In liberal democratic states governments and bureaucrats manage the demands of political parties, the media, competing social groups, and classes (along with international responsibilities, dynamics, and pressures) in the context of their own values and ideological framing. Thus the basic functional constraints identified by critical theorists do not appear as "objective constraints" but instead are filtered through the prism of different ideational frames by differently situated social actors within the state and civil society. While the outer limits of state policy making may still be understood as shaped by the strength of the economy, those limits are always "spongy" and contestable, as the *content* of policy can never be *reduced* to the impersonal dynamics of an economy. Constructivists would insist that interpretations or ideational renderings of these limits matter, not in the sense that they cause material change in any simple mechanical sense but rather in the sense that shifts in shared understandings provide the meaning and context in which shifts in material practices take place. In this respect ideas and values provide a narrative explanation of material change. As John Ruggie explains, a narrative explanation is not a deductive-nomological explanation of events; rather, it both "thickly" describes and configures events, rendering certain "facts" more significant than others and arranging or (after Polkinghorne) "emplotting" them into a more or less coherent gestalt.[36] Both the factual and normative content of such narratives can always be contested, but sometimes they are broadly accepted. For example, Alan Schnaiberg has persuasively argued that the "treadmill of production" in capitalist societies persists because of the persistence of a broad social consensus on the need for economic growth (rather than because of any functional imperative of the state). It is this broad consensus, and not merely any artful management on the part of state elites, that serves to render the contradictions mostly invisible, or at least something society must live with.[37] In this sense the functions of accumulation and legitimation are *both* discursively produced, and they can be discursively challenged and emplotted in different ways. Here I agree with Rosewarne that such an understanding "provides a different insight into the teleology of an ecologically bound capitalism."[38] Again, while I cannot deny the usefulness of higher order theoretical abstraction in shedding light on the inertia of existing social structures, I seek to show (in the final section of this chapter) the importance of combining the material and ideational levels of analysis of social life as a means of understanding the prospects for the emergence of greener states.

The functionalist theories of the capitalist welfare state, however, have highlighted the ways in which the outer boundaries of successful policies—green or otherwise—appear to be set by the economy. Nowadays it seems that everyone has become *resigned* to the ways in which market processes have increasingly disciplined political action within and beyond the state. Any orchestration of ecologically sustainable development by a regime of green taxes is quite likely to have the inflationary consequences predicted by O'Connor unless very carefully phased and managed.[39] In this sense Dryzek's reference to the imprisoned and unimprisoned zones of policy making in liberal polities still resonates.[40]

A critical constructivist understanding makes it possible to question this resignation. What is missing from the state theories examined so far, and even more crucial to understanding both political resignation and political protest in this era of globalization, are the ways in which new ideological framings of the role and rationale of the state have helped to facilitate changing forms of capital accumulation. The problem is not simply that the pressures of economic globalization make it increasingly difficult for states to manage their own societies, economies, and environments but also that there has been a palpable shift in the *dominant understanding* of the very role and rationale of the state in ways that appear to make it even more difficult to uphold the values of environmental protection and environmental justice *through* the state. Social aspirations and expectations of the state have changed in ways that make it harder to revive or reinvent an ethical ideal of the state as embodying any substantive social and ecological purpose. It follows that the discursive battle over the role and rationale of the state—and how it might respond to global competitive pressures—must form a crucial part of any understanding of, and political mobilization toward, an alternative, ecologically sustainable society. After all, it is these discursive contests that bring into view what "could be otherwise." As I seek to show below, the future of ecological modernization is uncertain and will ultimately be the outcome not of strategic management by state elites of objective imperatives but rather of political and discursive contests over the role of the state and the modernization process.[41]

3.3 From the Welfare State to the Competition State

So far I have explored critical theories of the capitalist welfare state that have focused primarily on its domestic role in managing the contradictory imperatives of accumulation and legitimation. Here I turn to the effect of intensified globalization on the prospects for the green state. That is, *to the extent* to which production has become more flexible, trade more open, capital more mobile, and financial markets more integrated, the competitive pressures of capital accumulation might be expected to increase and the political autonomy and legitimacy of the state to diminish accordingly.[42]

Within the discipline of critical international political economy increasing attention has been directed to the ways in which state policies (from defense and trade policy, to fiscal and monetary policy, to education, law and order, and the environment) are being defined in terms of comparative international competitiveness. In place of social democracy and the welfare state we now find the ideological ascendancy of neoliberalism and the competition state, whose primary task is to make economic activities located within the territory of the state more competitive in global terms.[43] While competitive pressures are certainly not new, the nature of the competitive game has changed.[44] Whereas John Ruggie had called the two-level process of globalization and state-welfarism in the post–World War II period "embedded liberalism," Philip Cerny has called the new game the "new embedded financial orthodoxy."[45] The rules of this new game require states to pursue a now familiar repertoire of measures that include reducing government spending, deregulating labor and financial markets, privatizing state-owned enterprises, dismantling protection and other trade restrictive measures, controlling inflation and sidelining macroeconomic demand management in favor of a microeconomic reform designed to improve international competitiveness. As Robert Cox explains, "neoliberalism is transforming states from being protective buffers between external economic forces and the domestic economy into agencies for adapting domestic economies to the exigencies of the global economy."[46] Fellow critical theorist Stephen Gill has called this "disciplinary neoliberalism" (which he traces to a particular Anglo-American model of capitalist development) that is increasing the power of investors relative to other members of civil society and thereby promoting "a social Darwinist reconfiguration of priorities, policies and outcomes."[47]

The upshot has been a shift away from the idea of the state as the protector and provider of public goods and services toward the notion of the state as a facilitator of privatization, commodification, marketization, and deregulation. The idea that governments should "steer and not row" has served as one influential neoliberal recasting of the *rationale* of government to that of coordinator rather than provider of social services, a shift that has also ushered in profound changes in public sector management.[48] Yet even the role of coordinator of service delivery remains secondary to the main strategic game of improving competitiveness.

There are two concerns for critical political ecologists arising from these developments. The first is that the systematic and widespread application of neoliberal policies by states can become self-fulfilling insofar as they work to strip states of their steering capacity and political autonomy. The second concern is that the quest by the competition state to attract capital and improve the competitiveness of the national economy exerts a downward pressure on domestic and global environmental standards, setting off "a race to the bottom." Such a race can only end when all states continue to reduce their environmental requirements until they are on a par with those in the "dirtiest states."

Turning to the first concern, it is clear that not all states are losing domestic political autonomy to the same degree. It is the unevenness of globalization that has generated debate among globalization scholars as to whether states are weakened victims or powerful conductors and beneficiaries of economic globalization. The debate remains alive because there can be no single answer to this question. The response can vary depending on the policy domain and the state or state grouping we might choose to examine, as there is any number of different ways of grouping states for such purposes (e.g., hegemonic or nonhegemonic; strong or weak; developed, developing, or underdeveloped).[49] At one extreme, we can expect developing and underdeveloped countries that are stricken with high levels of external indebtedness and beholden to the structural adjustment policies of the International Monetary Fund to be passive victims suffering a lessening of both political autonomy and perhaps legitimacy as economic globalization intensifies. At the other extreme, powerful states such as the United States and Germany are in a position to promote and shape the global or regional economic order in ways that suit their own strategic interests or their own particular understanding of what a just or secure world or regional order might be. However, in some policy domains, such as monetary or exchange rate policy, even powerful states can be subjected to swift punishment in international capital markets.[50]

In between these extremes, the disputes over the relationship between the state and globalization are most vigorous. Nonetheless, we can dismiss the pronouncements of hyperglobalizers such as Kenichi Ohmae that *all* states are passive victims of global change as this creates the misleading impression that *all* the significant changes are coming via decisions made beyond the state.[51] Such a casting of the problem obscures the fact that the processes of economic deregulation are *being actively orchestrated by particular states* through various multilateral arrangements.[52] Global markets cannot exist without the national legal systems of states, which provide the basic stability, contractual certainty, and the protection of private property rights necessary for investment.[53] Despite the huge transformations wrought by globalization, a significant subset of developed states remains the gatekeepers of the global order, the linchpin in the larger global governance network, creating and conferring power and legitimacy on other nodes of governance.[54] The *potential* power of states acting in concert to change these arrangements is clearly enormous.

However, the challenge of change is daunting given the path-dependency of policy regimes and the fact that not all states are able to play the role of gatekeepers of the global order. It is mainly the more powerful states in the OECD—above all the United States—that are the real conductors of globalization, while most developing states are increasingly victims rather than agents of economic globalization.[55] If change is to occur, a particularly important responsibility rests with middle power states to join with weakened states to challenge unjust global practices concerning trade, aid, technology transfer, and debt and remake the rules of global economic and environmental policy along green lines.

Moreover, to accept that most states have suffered some loss in political autonomy is not the same as saying political autonomy has come to an end or that the welfare state has become extinct. Geoffrey Garrett has persuasively shown that the evidence in both OECD and non-OECD countries on the relationship of globalization to domestic political conditions and economic policy belies the common argument that national autonomy is in decline.[56] Cerny's "new embedded financial orthodoxy" has not displaced the "embedded liberalism" of the previous era, although it has left it somewhat altered. State welfare services have become *more* rather than less important as a compensatory mechanism to protect the less privileged members of society from the harsher consequences of economic globalization. Likewise it can be argued that the environmental regulatory role of the state has become more rather than less important to protect ecosystems and life support systems from the intensified pressures of economic competition.

Turning to the evidence of downward pressure on environmental (and labor) standards in many industries and regions, what former US presidential candidate Ross Perot referred to as the "giant sucking sound" of firms and jobs moving south in the wake of NAFTA has its counterpart in increased pollution and disease in the *Maquiladora* region in Mexico. Firms operating in a relatively lax regulatory environment are often able to gain an unfair export price advantage vis-à-vis firms operating in more heavily regulated jurisdictions. Such practices are central to the success of many firms operating in free trade zones, where lax environmental and labor standards have enabled such firms to produce low-priced goods for export. These general competitive pressures are frequently invoked by industries to resist further environmental regulation by the state. Thus, regardless of whether the competitive pressures from offshore pollution havens are real or imagined, they can have a chastening effect on policy makers who respond in highly sensitive ways to industry complaints and threats to relocate.[57] It might also be plausibly argued that the only way to curb this downward race and minimize the competitive advantages gained by firms operating in pollution havens is for states collectively to pursue environmental harmonization, by negotiating multilateral environmental treaties that lay down collectively agreed environmental standards and requirements.

Yet for many reasons the downward pressure is not likely to culminate in an anti-ecological endgame. The reasons include resistance from civil society, changing regulatory practices on the part of states, and the adaptability of capitalism to changing state regulation and consumer demand. The finer-grained critical analyses of the competition state are increasingly emphasizing the diversity of state responses to competitive pressures. That is, there is no unitary model of the competition state that drives domestic regulation ineluctably downward. Different states seek to cope with change in different ways by pursuing different *competitive strategies* that seek to improve the climate for business in order to enhance national competitiveness in the global economy. While some of these strategies seek to deregulate, others seek to tighten regulations or simply regulate differently. Politically each state produces distinct "historical compromises" that shape the nature of the state's response to internal and external pressures.[58] As Ronan Palan and Jason Abbott explain, these "competitive strategies are shaped primarily by the constellation of interests within the state, and by the struggle for accommodation between them."[59] Understanding state responses in this way also directs attention to the socially contested rather than merely functionally driven character of the push toward greater national competitiveness.

One such strategy is ecological modernization, whereby states would seek to enhance the competitiveness of industry by unilaterally *increasing* rather than decreasing the stringency of environmental regulation. Far from resulting in economic contraction, such regulation has sometimes acted to further economic growth, particularly in environmental industries, while also creating an upward rather than downward ratcheting effect in environmental standards. The fact that those states with some of the strictest environmental regulations in the world, such as Germany and the Netherlands, also have strong economies certainly questions the global generalizability of the race-to-the-bottom argument.[60] In some industry sectors, the existence of a regulatory gap between developed and developing states has seen the "bottom rise" with economic growth.[61] According to Maarten Hajer, while the discourse of ecological modernization is not hegemonic (in the sense that it is the only environmental policy discourse), it has nonetheless become the most politically and economically credible way of talking about environmental policy since the 1980s.[62]

Of course, any shift toward more stringent environmental regulation by particular states will invariably affect some firms and industries more than others (forcing some to the point of closure), creating economic winners and losers as environmentally unfriendly subsidies are removed and new green taxes and regulations are imposed. The more rapid are such regulatory and fiscal shifts, the more inflationary they are likely to be. As such an outcome would underscore the political limits of any rapid, stringent and concerted greening of the economy, an argument for such change must include its careful phasing and management.[63]

3.4 Ecological Modernization: Just a New Competitive Strategy?

In view of the foregoing, what are the conditions and virtues of ecological modernization as a "new competitive strategy" for aspiring green states responding to the pressures of economic globalization in the new millennium? In particular, to what extent might such a green strategy alleviate the tension between the state's accumulation and legitimation imperatives? To what extent, and how, might ecological modernization free the hand of the political leaders of aspiring green states to orchestrate more systematic environmental protection and environmental justice?

In order to address these questions it is necessary to clarify the range of practices and normative understandings of ecological modernization. Peter Christoff, for example, has distinguished a relatively simple (uncritical) understanding of ecological modernization, which amounts to little more than a cost minimization strategy for industry, from a more reflexive (critical) understanding that seek deeper structural transformations in the economy, society, and the state, and encompasses a precautionary approach to risk assessment and consensual/democratic policy-making styles.[64] The process of reflexive modernization entails not simply the more efficient pursuit of goals but critical reflection on the goals as well. "Reflexivity," according to Anthony Giddens, "here refers to the use of information about the conditions of activity as a means of regularly reordering and redefining what that activity is."[65]

These contrasting approaches may be understood not as mutually exclusive but rather as part of a continuum of possibilities defined by weak and strong poles.[66] As these different possibilities are laid bare, it should be clear that weak ecological modernization offers, at best, only a short-term reprieve to the tensions between accumulation and legitimation facing the state. That is, weak ecological modernization can only slow down the rate of ecological deterioration, but it cannot reduce aggregate levels of environmental damage and therefore cannot resolve the contradictions of the capitalist state over the long term. Reflexive modernization, in contrast, calls into question the policy tools and goals as well as the purpose and meaning of the modernization process. Reflexive modernization is no longer *just* a competitive strategy, it becomes a means of economic and societal transformation.

Even at its weakest form, ecological modernization challenges the traditional view that too many environmental restrictions, taxes, and costs would make industry less competitive, leading to a slowdown in economic growth rates, unemployment, and possibly capital flight. It also challenges the traditional idea of environmental policy as mere damage control. Appearing in the early 1970s as an afterthought to other policy programs and state infrastructure, environmental policy programs tended to be reactive and remedial, dealing with environmental damage, much of which had been directly or indirectly sponsored by other agencies of the state, such as those managing agriculture, industry, and trade.[67] The afterthought or add-on status of environmental policy was also reflected in environmental budgetary outlays, which tended to form a very small percentage of state budgets. Against this background, the pioneering analyses of the contradictory functions of the welfare state in the 1970s more or less coincided with the orthodox thinking of the time. That is, neo-Marxists, state elites, and environmentalists (both radical and reformist) recognized that environmental problems could, at best, be politically managed by governments artfully balancing conflicting economic and environmental interests. However, for critical theorists, environmental problems could not be fully resolved so long as the institutions of capitalist societies remained intact. For radical environmentalists, such an analysis provided all the more reason to call into question capitalist society.

However, from the 1980s and beyond, the zero-sum assumptions concerning the relationship between economic growth and environmental protection began to be challenged by the new discourse of ecological modernization. The basic idea that economic and environmental considerations could be made to work synergistically became central to the more diffuse policy discourse of sustainable development, which achieved world prominence following publication of *Our Common Future* by the World Commission on Environment and Development (WCED) in 1987 (widely known as The Brundtland Report).[68] Indeed, Maarten Hajer has described The Brundtland Report "as one of the paradigm statements of ecological modernization," with the sustainable development debate serving as the "first global discourse-coalition in environmental politics."[69] The good economic news offered by proponents of ecological modernization, reinforced by The Brundtland Report, was that environmental protection can, under the appropriate conditions, act as a spur to further technological innovation, more economic growth and higher levels of prosperity.

Whereas ecosurvivalists of the 1970s, such as William Ophuls, had raised the specter of the state as an all powerful environmental Leviathan, ecological modernizers presented a much more benign portrait of the state as an agent of ongoing modernization.[70] On this new view, the role of the state switched from being simply reactive, remedial, and/or punitive to being proactive, anticipatory, and preventative, or as Mol has put it, "from *dirigiste* to *contextual* 'steering.'"[71] Such a portrait of the state not only appeared to coincide with the increasingly influential governing philosophy of "steering, not rowing" in the new era of fiscal austerity but also served the competition state's main strategic game: improving economic competitiveness. This simple win-win approach has been aptly called "strategic environmental policy."[72]

At the core of this strategic case for ecological modernization is the notion that economic growth and environmental deterioration can be decoupled by pursuing greener growth rather than by slowing growth. By "greener growth" it is usually meant economic growth that uses less energy and resources, produces less waste per unit of gross domestic product, and seeks constant technological innovation in production methods and product design in ways that are less material-energy intensive. Ideally this process points toward the development of closed-loop production where nothing is wasted and everything is reused or recycled. Such improvements on any one or more of these fronts signal improvements in eco-efficiency or environmental productivity. In a far cry from the original doomsday report to the Club of Rome and as a tribute to the virtues of eco-efficiency, the new report to the Club of Rome, called *Factor Four*, now maintained that wealth would likely double if resource use could be halved.[73]

Whereas advocates of limits to growth had sought a far-reaching transformation of industrial society (indeed, some have suggested a complete about-face)[74]—a demanding claim that required extensive societal deliberation about the shape of our collective future—ecological modernization understood merely as a process of continually improving environmental productivity by means of new technologies and management practices required no such societal debate. By naturalizing rather than questioning the process of modernization via technological change, it was consistent with ongoing de-traditionalization, ongoing individualization, and increasingly sophisticated private consumption patterns. It is precisely this lack of any deep critique that explains the distinction between *simple* versus *reflexive* modernization.[75] As Ingolfur Blühdorn puts it, the eco-efficiency movement has transformed and neutralized environmentalism from the "flight from technology" to a "technological attack," without making explicit and defending the values that ecological modernization is supposed to promote.[76] In short, Blühdorn maintains that such technological modernization merely "restores the illusion of rational progress and control."[77] Michael Jacobs, in an influential Fabian pamphlet entitled *Environmental Modernization: The New Labour Agenda*, has sought to exploit the technical discourse by recommending what he calls "environmental modernization" to New Labour in Britain precisely because it is not a value-driven discourse and it is therefore not necessary for one to adhere to green ideology to support ecological modernization.[78] The same might be said for the zealous pursuit of ecological modernization since 1996 by Sweden's ruling Social Democratic Party, under the leadership of Prime Minister Göran Persson. In building a broader societal coalition to green Swedish industry, and turn the Swedish welfare state into a *green* welfare state, Persson has skillfully exploited the strategic advantages of ecological modernization in ways that have put both the Green and Left parties on the defensive.[79] The Swedish Social Democrats can nonetheless be credited with having introduced some of the most far-reaching strategic, legal, administrative, and tax reforms to integrate environmental and economic policy in the world today.

Yet ecological modernization can masquarade worldwide as an ideology-free zone only for as long as it can succeed in "naturalizing" the modernization process to protect it from deeper questioning and social unrest. To the extent that serious ecological problems persist and the distribution of environmental harm remains highly skewed, then, despite ongoing technological innovation, one can expect public anxiety to grow, public trust in experts and managers to wane, and political agitation or disaffection within civil society to grow. To the extent to which this occurs (both domestically and globally), ecological modernization understood as mere "technological adjustment" can no longer be insulated from further confrontation and critical questioning. Quite the contrary, increasing reflexivity necessarily calls into question the very processes of technological innovation and capitalist economic modernization, along with the interests and ends it serves. One can compare the mostly technical discourse of ecological modernization with the principles and recommendations of the Brundtland Report. Although the Brundtland Report stressed the economic advantages of sustainable development, it did not rest its case primarily on economic arguments alone. By defending sustainable development in terms of balancing the principles of intra- and intergenerational equity, the Brundtland Committee ultimately offered a future-oriented, human-centered environmental justice argument that rested on the fulfillment of all human *needs*, now and in the future.

In contrast, the technical case for ecological modernization is primarily concerned with means (how to pursue greener growth) rather than with ultimate ends. As Christoff has argued, it is concerned with mere "technological adjustments" toward greater eco-efficiency at the level of the firm. This is essentially an economistic understanding that does not challenge existing institutions or dominant neoliberal economic policies.

Moreover as a technical discourse it may be easily administered by technocratic policy makers employing the traditional regulatory and fiscal policy levers of the nation state. Such an approach might readily be adopted and pursued more systematically as an adaptive strategy to economic globalization by existing capitalist states. This is why it can be recommended as comfortably fitting into the policy goals of New Labour, thereby providing an "environmental Third Way."

Since this eco-efficiency strategy seeks only to "pick the low-hanging fruit" offered by the ecological modernization discourse, one might assume there to be few obstacles to its incorporation into the policy platforms of most political parties. Yet even this relatively straightforward political task demands more ecologically sensitive accounting techniques and management practices, tighter policy coordination and integration, and some modest realignment in state goals and practices to bring environmental concerns to the fore in fiscal and industry policy. To coordinate the changes, governments would face some tough battles with industries that have been beneficiaries of ecologically damaging subsidies (e.g., agricultural subsidies or low-energy charges relative to other industries) or other government assistance. Without the political mobilization of the radical ecology movement, the momentum for ecological modernization would slow down and possibly come to halt once all the accessible fruit is plucked. In this respect Andrew Dobson has rightly concluded that ecological modernization would not be possible without the ongoing educative role played by radical environmentalists; as he explains, "Radical ecology's role for the twenty-first century is as a condition for the possibility of its reformist cousin."[80]

Strong ecological modernization challenges each and every assumption of its weak eco-efficiency counterpart. As Christoff shows, the weak interpretation may be contrasted with an ecological interpretation that places at center stage the ecological integrity of ecosystems and life-support systems, the Eurocentric orientation of the conventional discourse of ecological modernization may be contrasted with an international perspective, the conventional focus on pollution control and clean production may be expanded to encompass biodiversity and wilderness preservation, the simple process of technological adjustment may be expanded into multiple pathways to ecological modernization, and technocratic management may be opened into the discursive democratic negotiation of policy and institutional change.[81]

The main objection to weak ecological modernization is that improvements in eco-efficiency alone, while welcome, do not necessarily lead to a decrease in environmental degradation in *absolute* terms, especially when examined from a global rather than national-point of view, and they do not necessarily address global environmental injustices. As Jacobs has conceded, absolute growth in global environmental degradation far exceeds improvements in environmental productivity or eco-efficiency.[82] While it is always possible to improve eco-efficiency, it is not possible to achieve 100 percent eco-efficiency and as world production, trade, and population continue to increase in absolute terms, overall environmental degradation will inevitably continue to rise (albeit at a diminishing rate). This serves to underscore the somewhat paradoxical relationship between economic growth and environmentalism.[83] That is, growth not only increases environmental degradation but also gives rise to a greater societal capacity to respond to such problems by pursuing more ecologically efficient growth.

The problem, however, is not only that the pursuit of eco-efficiency is merely a short-term solution to environmental degradation. It is also that the capacity to pursue even this limited strategy is not uniformly enjoyed by all states. Christoff, in particular, has queried whether a strategy of weak ecological modernization can be pursued by developing states dependent on commodity exports and lacking in the basic infrastracture, resources, and know-how to pursue successful environmental policies. Clearly, more concerted assistance by the developed world toward capacity building in the developing world is essential if the short-term benefits of weak ecological modernization are to be shared.

The discourse of ecological modernization has centered more on industry policy—on the minimization of material use, energy intensity, and waste production—than on the full range of environmental protection concerns. Other environment protection measures, such as biodiversity preservation and wilderness protection, place constraints on economic growth, while still others curb growth altogether, at least by conventional measures. Ecological modernization may to some extent reduce the tensions between economic growth and environmental protection in some industry domains, but it has not resolved the tensions in all domains. In many policy domains difficult political trade-offs remain to be made.

Not all industries can enjoy the competitive advantages that flow from being the first mover in technological innovation. Not all industries can play follow the green leader either to the same degree. Many industries are inherently environmentally degrading, such as the nuclear and fossil fuel industries, and therefore must be phased out rather than merely modernized. The need then is for major economic restructuring through policies that can ease the transition away from ecologically damaging industries to ecologically benign alternatives.[84]

More important is the issue of the OECD countries achieving their environmental improvements by the relocation of dirty or energy-intensive industries to developing countries, as local standards have raised the costs of doing dirty business. Polluting industries cannot forever be moved around the globe.[85] Avoiding the relocation of dirty industry to the South requires, for example, green technology transfer to enable developing countries to leapfrog over the ecological mistakes committed by the developed world. Yet to restrict such a strategy to mere technical fixes would be to assume a naïve, gradualist, evolutionary view of modernization that underplays social conflict and the discursive struggles associated with environmental and economic policy making. Ultimately, from the perspective of strong ecological modernization, such problems can only be addressed by tackling the structural inequalities in the global economic order that impede the ability of many states to pursue stronger domestic strategies of ecological modernization. Increasingly reflexive approaches to ecological modernization acknowledge that there is not only one path to prosperity, that prosperity need not mean high-consumption lifestyles, and that the problem-solving capacity of states and societies is partly a function of the ability of their social institutions to respond to both positive and negative feedback from society and the environment.

Ecological modernization clearly has a continuum of meanings and political possibilities, and it can take both weak and strong directions. Advocates of strong ecological modernization accept the necessity of more ecologically friendly technologies but insist that such technologies are not sufficient for addressing ecological problems.

As Douglas Torgerson explains, this more discursive expression of ecological modernization involves both functional *and* constitutive aspects of green politics.[86] Instead of merely ironing out the dysfunctional elements of capitalism and the administrative sphere to ensure that they function in ways that are more ecologically rational, these more radical proponents of ecological modernization throw into question both means and ends; ". . . modernity appears flawed and incomplete because by its own standard of rationality the irresistible momentum of progress rules out rational deliberation over its purpose and direction."[87]

Indeed, this problem amounts to an ambivalence toward ecological modernization by so many green theorists. As Paul Rutherford explains, this ambivalence "reflects the fundamental philosophical dilemma of the *dialectic of enlightenment* in which modernity, with its dependence on rational-instrumental, scientific knowledge, embodies a self-destructive social relation to nature."[88] Hajer is equally ambivalent on this score, vascillating between seeing ecological modernization as either mere green window dressing for business as usual or a subversive discourse that may serve as "the first step on the bridge that leads to a new sort of sustainable society."[89]

However, from the perspective of critical political ecology, this tension between the functional and constitutive dimensions of ecological modernization, which represents a vacillation between functional adaptation and systematic negation of the established order, carries the potential to transform or reconstitute the institutional and societal self-understanding of the ends and means of social life in emancipatory ways. Growing public anxiety about the scale and gravity of ecological risks has resulted in a loss of *trust* in scientific and technocratic experts and state elites in managing ecological problems.[90] Similarly the increasing temporal and spatial reach of ecological risks has spawned critical reflection on the short-time horizons and territorially contained character of our preeminent political institutions of ecological management—states. The environmental justice movement, in particular, has drawn attention to the unfair distribution of ecological risks between different classes and regions. At the same time developing countries have accused developed countries of creating a larger "ecological footprint"[91] relative to developing countries while denying the developing world the same rapid resource exploitative path to economic prosperity. Finally, the commodifying and disembedding dynamic of modernisation—which has lifted social relations out of traditional and local structures and rearranged them across more regional and global space-times—has prompted renewed efforts to defend local biodiversity, places, habitats, and vernacular cultures from economic commodification and cultural appropriation.[92] There is, then, significant potential for a wide movement toward stronger ecological modernization. And this is especially so given that, as Robert Paehlke puts it, "Green ideas are one of the few perspectives other than neoliberalism that exist in most nations."[93]

More reflexive modernization is neither antimodern nor postmodern but rather entails the *radicalization* of modernity.[94] The radical ecological critique of modernity is not a premodern critique, since it is dependent on scientific research, new technologies, and new forms of communication, which it then enlists to challenge the idea that ecological problems can be solved merely by ongoing technological adjustment.[95]

The process of critical reflection that is central to the method of immanent critique is also central to the processes of reflexive modernization. For Ulrich Beck, reflexivity forces a *self-confrontation* by industrial society with ecological consequences that cannot be addressed by the mere perpetuation of existing ideas, institutions, and practices of governance.[96] It is this very practice of self-confrontation that leads to *critical reflection* over the condundrum of accountability (or lack thereof) for the proliferation of ecological risks and hazards, which in turn leads to a redefinition of responsibility, safety, liability and risk distribution. Indeed, Beck has called reflexive modernization the "second Enlightenment,"[97] or the "ecological enlightenment,"[98] since it calls for a radical rethinking of the relations of definition, understood as the structures of authority that define, assess, and manage risks, and, arising out of this rethinking, the development of new forms of ecological democracy.

3.5 Globalization, Sustainability, and the State

Environmental protection emerged as an additional, identifiable, but subsidiary task of the welfare state in the 1970s, to be later heralded in the 1980s and 1990s as a more significant (although still not primary) task to be integrated with, rather than overshadowed by, the goals of economic development.[99] However, the "win-win" strategy of weak ecological modernization as mere technological adjustment can merely offer advances in environmental productivity; it cannot cover all the fields of environmental concern in the modern risk society and it ultimately cannot head off further environmental degradation in the longer run. More reflexive modernization is clearly needed. But how much more reflexive must that be?

State and societal learning may range from a simple adjustment of policy tools (which may uncritically accommodate deeper contradictions in the broader policy environment) to an innovative transformation of policy direction in response to a critical analysis of existing structures, past failures, new circumstances and new knowledge culminating in fundamental institutional transformation. To highlight the ways in which reflexivity works in degrees, it is useful to distinguish (building on the work of P. A. Hall and Norman Vig)[100] at least four different levels of state reflexivity:

- Change in policy instruments
- Change in policy goals
- Change in policy paradigm, or the hierarchy of policy goals
- Change in the role of the state

According to Norman Vig, changing the suite of policy instruments may contribute to policy learning but not necessarily to *policy innovation* "unless the use of these instruments is accompanied by changes in higher order principles."[101]

These levels correspond with a movement from less reflexive to more reflexive learning, culminating in critical reflection on policy tools, policy settings, the hierarchy of goals, and the *broader institutional context* in which policies are made and implemented. *Strong* ecological modernization discourse must therefore include *all four* of these levels of reflexivity. Strong ecological modernization has raised issues of new technologies, new instruments of environmental management, new policy settings and new policy principles (e.g., the polluter pays principles and the precautionary principle), new policy paradigms based on new societal goals (sustainable development), and new understandings of the role and rationale of the state and the state system (from environmental exploiter and facilitator of private environmental exploitation to public environmental trustee).[102]

The higher order, structural dimensions of this discourse may be seen as offering immanent critiques of lower order understandings. For example, the introduction of new policy tools such as the replacement of prescriptive regulation with market-based policy instruments, without a systematic critical analysis of the reasons for regulatory failure, will not necessarily lead to improvements in environmental outcomes, although economic efficiencies may result. Already regulatory failure has occurred because of inadequate resourcing, staffing, and monitoring by the relevant agencies and not necessarily because of any inherent defect in prescriptive regulation. However, if such failure persists despite proper resourcing and monitoring, then it is necessary to call in question broader policy strategies, and possibly policy hierarchies and the role of the state.

These different levels of reflexivity also shed light on the functionalist theories of the state examined earlier in this chapter. The policy limits and policy failures identified by these theories point to contradictions in the social structure of capitalist societies, which encompasses the capitalist economy and the liberal capitalist state. Translated into the different levels of analysis, the argument is that without a transformation in the role and functions of the liberal capitalist state and capitalist society (level 4), the contradictory requirements of legitimation and accumulation cannot be solved (as distinct from politically managed) by changes in the policy tools, policy setting, and or even the hierarchy of policy goals. It is the character of the system (the mutual dependencies of the capitalist economy and the liberal capitalist state) that that set limits to the effectiveness of such state interventions. A deep and lasting resolution to ecological problems can therefore only be anticipated in a postcapitalist economy and postliberal democratic state.

To date, however, the paradigm shift in most OECD countries has been mostly confined to levels one and two, and only falteringly three. For example, in a recent comparison of environmental policy developments in the United States and the European Union, Norman Vig has identified what he calls a paradigm shift in regulatory principles involving "an array of neoliberal strategies and instruments with the general philosophy of sustainable development."[103] This peculiar mix of new market-based strategies (which can be traced to broader shifts in national and global economic policy directions) with the new discourse of sustainable development incorporates both potential synergies *and* tensions. The higher order ideological shifts in government *economic* policy (which must also be located in the context of the intensification of economic globalization) still appear to be exerting much stronger pressure on the direction of environmental policy than the subversive dimensions of reflexive ecological modernization. These observations apply at both the national and international levels of environmental policy.[104] As noted, these shifts in economic thought and practice are, in turn, changing public understandings about what can be expected of states in an era of rapid globalization.

Usually states are not the first places where one might expect to find any significant motivation to reorder the hierarchy of economic and environmental policy goals, including corresponding bureaucratic and ministerial hierarchies, least of all any transformation in the role and rationale of states as institutions of governance. The source of normative-cognitive innovation is more typically found in local, national, and transnational environmental organizations and other advocacy coalitions, policy professionals and scientists, universities and think tanks, local networks and communities, progressive business, and international organizations and multilateral arrangements. In this respect the higher order normative-cognitive shifts are more likely to *culminate* in rather than *originate* with the state, as a result of a long, extensive, and mutually reinforcing series of changes at multiple levels, including civil society and the economy. This culmination is more likely to occur as more environmental NGOs and other environmentally concerned nonstate actors seek to transform the state in ways that enable it to "act back" on society in pursuit of greener goals.

Changes in the role and rationale of states also need to be accompanied by improvements in environmental capacity, which the OECD has defined as "a society's ability to identify and solve environmental problems."[105] As Jänicke and Weidner have shown, environmental capacity is not just restricted to government policy. Rather, it refers to the structural preconditions for societal solutions to ecological problems, including ecological, technological, and administrative knowledge, legal and material resources, policy institutions, political participation, and the strength of environmental organizations relative to opposing economic interests.[106] On this view, *environmental capacity building* may also be seen as part of the dynamic process of ecological modernization—a process that points toward increasingly reflexive modernization, since it represents the "institutionalisation and internalisation of new stages of problem-solving capacities in reaction to (or anticipation of) societal challenges or crises."[107]

Most states are a long way from this green ideal, even in Western Europe. Nonetheless, Hajer has claimed that "the practice of ecological modernization is undoubtedly the now dominant response of contemporary European societies to the so-called environmental challenge."[108] In contrast, Golub has argued that the EU still relies to a considerable extent on the old carrot-and-stick approach, although he concedes that there are also many instances where the Commission has pursued such win-win policies.[109] It is beyond the scope of this book to conduct the necessary empirical inquiry into ecological modernization in different states; my concern in this chapter has merely been to explore critically the arguments and theoretical insights emerging from the literature. I have sought to show that while there are many *economic* opportunities for pursuing weak ecological modernization by advanced capitalist states, such strategic environmental policy cannot tackle *all* ecological problems. Economic competitiveness, after all, is not an end in itself.

To the extent that stronger forms of ecological modernization may take hold, we should expect to see more reflexive (and hence more democratic) states that might also assume the role of ecological trustee and quell the growing public anxiety about ecological risks. Would a full-fledged green democratic state still be a capitalist state? On the one hand, the green state would still be dependent on the wealth produced by private capital accumulation to fund, via taxation, its programs and in this sense would still be a capitalist state. On the other, securing private capital accumulation would no longer be the defining feature or primary raison d'être of the state. The state would be more reflexive and market activity would be disciplined, and in some cases curtailed, by social and ecological norms. The purpose and character of the state would be enlarged and therefore be different. In this respect the green democratic state may be understood as a postcapitalist state. In the next chapters I explore what institutional reflexivity might entail in terms of a postliberal ecological understanding of democracy.

Of course, the capacity to pursue ecological modernization varies from state to state. Whether as strategic environmental policy or as the basis for far-reaching societal transformation, ecological modernization is a luxury that only a few privileged Western states are currently in a position to pursue in any systematic way. This is not an acceptable situation in the long run, and it can only be defensible in the short run if those states that currently pursue ecological modernization deploy their "green wealth" to further environmental and social justice goals that may not be so easily harmonized with national economic pressures. There should be positive spin-offs for global society to the extent that the privileged green states are able develop greater institutional reflexivity of a kind that is more sensitive to global environmental protection and global environmental justice. We would also expect such states to be in the best (relative) position to act as good international citizens, whether unilaterally (by offering more reflexive environmental and economic policy discourses and more ecologically reflexive domestic institutions for emulation by other states) or multilaterally (in setting the pace in difficult multilateral environmental negotiations).

4

The Limits of the Liberal Democratic State

4.1 The Liberal Democratic State: Not Reflexive Enough?

The modest greening of multilateralism and capitalism are the products of policy learning on the part of states as well as nonstate actors. However, more reflexive learning requires a free and critical communicative context in which wealth and risk production and distribution decisions take place. This chapter asks whether the liberal democratic state—heralded by some as the best and final form of government in modern times[1]—has the reflexive learning capacity to usher in stronger forms of ecological modernization and more environmentally just forms of risk assessment.

It may be argued that the liberal democratic state already possesses the requisite adaptive learning capacity to tackle ecological problems—a claim that finds empirical support in the relatively superior environmental record of liberal democratic states when compared to single-party communist states, particularly in the decades preceding the fall of the iron curtain. The reason for this is that liberal democratic states have provided far more scope than totalitiarian states for the public exposure of ecological problems and the political mobilization of environmental concerns, ranging from mass protests, to citizens' initiatives, to the formation of environmental organizations and green political parties. Such environmental activism on the part of citizens and nongovernment organizations has enabled the generation of ecological information and critical publicity that has, in turn, helped prompt environmental technological innovations, improve the responsiveness of the liberal democratic state, and improve environmental outcomes.[2]

My concern in this chapter is not to deny that the liberal democratic state is reflexive but rather to ask whether it is *reflexive enough* in moving toward more ecologically sustainable societies. In this respect the superiority of liberal democratic states over totalitarian ones ought not to serve as an ecological vindication of liberal democracy or of capitalism. This chapter seeks to build upon a growing body of work by green political theorists that challenges the capacity of the liberal democratic state to resolve, as distinct from manage, ecological problems. In so doing, this chapter confronts the third major green hesitation toward the project of building greener democratic states.

There are two, analytically distinct, dimensions of this green critique of the liberal democratic state, although in practice they often converge. The first dimension concerns the *state form* through which liberal democratic ideals have been institutionalized. While eco-anarchists have historically led this charge against the liberal democratic state, they find support from poststructuralists concerned with the exercise of biopower and from critical green theorists who are troubled by the ascendancy of the administrative state, or what Douglas Torgerson has called "the administrative mind," over civil society.[3] Together, these critical green voices have variously maintained that all states, whether liberal or nonliberal, are in certain respects *inherently* unreflexive structures of governance precisely because they are coercive, highly centralized, and bureaucratic centers of power with a set of security, disciplinary, surveillance, and/or administrative imperatives that are fundamentally at odds with the green vision of participatory democracy and the ideal of the green public sphere. According to Torgerson, the green public sphere is ideally a *decentered* arena of debate, taking its place among a plurality of public spheres, where there is no group controlling or providing authoritative direction from any centre and no central agent of change.[4] Later, in chapter 6, I will strongly endorse this ideal of the green public sphere. In this chapter I seek to defend the state (in the form of a green democratic state) not simply as a necessary evil but rather as a crucial mechanism for facilitating democratic negotiations in the public sphere, and for steering society along more ecologically sustainable lines.

The second dimension of the green critique of the liberal democratic state relates to the *liberal* character of the liberal democratic state, which is shown to thwart the development of a genuinely public morality and associated notions of collective interests. This is the main argument I seek to develop in this chapter. There is now an extensive and growing body of green political scholarship that argues that liberal democracy is not especially conducive to protecting long-range, public environmental interests (e.g., biodiversity and ecosystem integrity).[5] This green critique also enlists and builds upon the longstanding critique of liberal democracy waged by social democrats, democratic socialists, and feminists to the effect that the class and gender inequalities generated by capitalism systematically undermine the conditions for the full enjoyment by all citizens of the political equality promised by the liberal democratic state. Thus the green critique adds further weight to the argument that the promise of liberal democracy is a false promise; while proclaiming to be universal, liberal democracy can be shown to be exclusionary in a variety of ways. Green political theory's unique addition to this longstanding critique is to draw out the links between democracy and environmental justice and to extend our understanding of the category of subjects excluded from any meaningful representation or participation in the liberal state, even though they may be harmed by decisions and actions made in the name of the state. In drawing on and developing the existing green critiques of liberal democracy, the last section of this chapter underscores the immanent character of the ecological critique of liberal democracy, and how this might lead to a reinterpretation of the meaning of self-rule in an ecological context.

These two dimensions of the green critique of the liberal democratic state, which focus on the respective democratic deficits of liberal democracy, and of the state form through which it is expressed, might be understood as mapping onto the informal boundary between the demand side and supply side (or democratic will formation and democratic will execution) of the ecological challenge from the point of view of the state. The boundary is an artificial one, not least because functionaries working within the administrative apparatus of the state are often routinely involved in both making and interpreting policies and legal rules. However, the boundary does help to elucidate what Martin Jänicke has described as "a steady deterioration of the *control ratio* between politics and the machinery of government."[6] That is, there is a growing disjuncture between those who make decisions, those who are politically responsible for them, and those who are affected by them. The influence of the democratically elected legislature over the state administration must pass through what he calls "the needle eye of ministerial responsibility."[7] Extending this critique of accountability, Ulrich Beck has attacked the state based administrative and legal system as a site of "organised irresponsibility" when it comes to managing ecological risks and harms.[8] Indeed, one of the core points of the more general green critique of the administrative state is that it makes something of a mockery of the liberal democratic ideal of public accountability. I therefore begin with the green critique of the supply side of the liberal democratic state, before moving on to consider the demand side.

4.2 The Ecological Critique of the Administrative State

The ecological critique of the administrative state takes seriously the proposition (explored in chapter 2) that the state, whether liberal or nonliberal, may not be the type of entity that is capable of *systematically* prioritizing the achievement of sustainability. This claim resonates with the realist view of the state system, which maintains that the anarchic structure of the state system makes questions of security, economic growth, and the competitive struggle of "staying afloat in a hostile world" the overriding preoccupation of states, with the consequences that environmental protection will forever remain peripheral (unless directly related to traditional questions of national security). In this section I focus on the domestic dimensions of this argument, which are directed to the unique modality of the organized and coercive political power of the state, and the distinctive capacities and motivations of state elites and managers (rather than the clash of social forces within civil society and/or the broader processes of societal modernization).

The idea that there may be *state* limits to democracy, and therefore to sustainability, is one that finds some support among non-Marxist theorists of the state who have focused on what they see as the essential, constitutive features of the modern state. According to the "organizational realist" approach of Theda Skocpol, states have their own unique organizational interests that cannot be simply explained in terms of social struggles in society, the capitalist mode of production, or capitalist society. Whatever else states may be or do, they are necessarily always "actual organisations controlling (or attempting to control) territories and people."[9] Accordingly she suggests that any theory of the state must be open to the possibility of the autonomous state as a source of power that is independent from society or capitalism. For Skocpol, this power resides in the tasks that are *uniquely* performed by states: resource extraction (i.e., taxation), administration and coercive control.[10] Resource extraction and maintenance of administrative capacity are required for internal order and security as well as external security and competition with other states.

Although Skocpol does not address green concerns, a similar organizational realist perspective is discernible in eco-anarchist perspectives on the state as well as green critiques of the administrative state. Following Skocpol, these green critics take seriously the fact that states are organizations that control (or attempt to control) territories and people. Alan Carter has argued that it was the failure of Marxist revolutionaries to recognize the autonomy of the state that led to totalitarian socialist states in Russia, China, and many developing countries.[11] According to Carter's state-primary theory, the nature of the relations of production that prevail in a particular society—as well as ecological problems—can be explained primarily in terms of *state interests* (particularly military interests), and not the interests of capitalist classes.[12]

In maintaining and developing their coercive powers, states have—over the centuries—sought ways of extending their revenue base, enlarging their sphere of influence within civil society and expanding and rationalizing their administrative apparatuses. The result has been the development of a centralized and hierarchical system of depersonalized and increasingly specialized bureaucratic power. In response to these broad historical developments, eco-anarchists and other critics of centralized state power have maintained that the "pseudorepresentative" administrative state, with its promotion of inegalitarian economic relations and "nonconvivial technologies," produces an environmentally hazardous dynamic.[13]

Green poststructuralists have likewise sought to deconstruct the disciplinary effects of biopower and green governmentality, while green critics of technocracy have lamented the cult of the expert, the so-called the scientization of politics, and the concomitant disenfranchisement of the lay public and vernacular knowledge in affairs of state administration.[14] The bureaucratic rationality of the administrative state is seen as too rigid, hierarchical, and limited to deal with the variability, nonreducability, and complexity of ecological problems.[15] Bureaucratic rationality responds to complex problems by breaking them down, compartmentalizing them, and assigning them to different agencies that respond to a hierarchical chain of command. This often leads to the routine displacement of problems across bureaucratic system boundaries.[16] Once we add to these developments the more recent revolution in public sector management, we have good reasons to concur with Paul Hirst that the traditional liberal architecture has increasingly "become a gross misdescription of the structure of modern societies."[17]

The tenuous link between popular political participation and control and technocratic state administration has also been a major theme in the work of Ulrich Beck. Indeed, Beck (like Martin Jänicke) argues that politicians and state functionaries act in ways that seek to mask problems rather than solve them. Ecological problems persist because they are generated by the same economic, scientific, and political institutions that are called upon to solve them. While the state cannot but acknowledge the ecological crisis, it nonetheless continues to function *as if* it were not present by denying, downplaying, and naturalizing ecological problems and declining to connect such problems with the basic structure and dynamics of economic and bureaucratic rationality. According to Beck, this organized irresponsibility can sometimes take on a Kafkaesque form. The state seeks to manufacture security by providing social insurance systems—health services, unemployment benefits, pensions, and workers compensation—but it can provide no protection against major hazards that can pierce the thin veneer of normality and expose the inadequacies of the welfare state. As Beck puts it: "What good is a legal system which prosecutes technically manageable small risks, but legalises large scale hazards on the strength of its authority, foisting them on everyone, including even those multitudes who still resist them?"[18]

It might be tempting to conclude from this general critique that states are part of the problem rather than the solution to ecological degradation. With its roots in the peace and antinuclear movements, the green movement has long been critical of the coercive modality of state power—including the state-military-industrial complex—and might therefore be understandably sceptical toward the very possibility of reforming or transforming states into more democratic and ecologically responsive structures of government. The notion that the state might come to represent an ecological savior and trustee appears both fanciful and dangerous rather than empowering.

Yet such an anti-statist posture cannot withstand critical scrutiny from a critical ecological perspective. The problem seems to be that while states have been associated with violence, insecurity, bureaucratic domination, injustice, and ecological degradation, there is no reason to assume that any alternatives we might imagine or develop will necessarily be free of, or less burdened by, such problems. As Hedley Bull warns, violence, insecurity, injustice, and ecological degradation *pre-date* the state system, and we cannot rule out the possibility that they are likely to survive the demise of the state system, regardless of what new political structures may arise.[19] Now it could be plausibly argued that these problems might be *lessened* under a more democratic and possibly decentralized global political architecture (as bioregionalists and other green decentralists have argued). However, there is no basis upon which to assume that they will be lessened any more than under *a more deeply democratized state system*. Given the seriousness and urgency of many ecological problems (e.g., global warming), building on the state governance structures that already exist seems to be a more fruitful path to take than any attempt to move beyond or around states in the quest for environmental sustainability.[20] Moreover, as a matter of principle, it can be argued that environmental benefits are public goods that ought best be managed by democratically organized public power, and not by private power.[21] Such an approach is consistent with critical theory's concern to work creatively with current historical practices and associated understandings rather than fashion utopias that have no purchase on such practices and understandings. In short, there is more mileage to be gained by enlisting and creatively developing the existing norms, rules, and practices of state governance in ways that make state power more democratically and ecologically accountable than designing a new architecture of global governance *de novo* (a daunting and despairing proposition).

Skeptics should take heart from the fact that the organized coercive power of democratic states is not a totally untamed power, insofar as such power must be exercised according to the rule of law and principles of democratic oversight. This is not to deny that state power can sometimes be seriously abused (e.g., by the police or national intelligence agencies). Rather, it is merely to argue that such powers are not unlimited and beyond democratic control and redress. The focus of critical ecological attention should therefore be on how effective this control and redress has been, and how it might be strengthened.

The same argument may be extended to the bureaucratic arm of the state. In liberal democratic states, with the gradual enlargement, specialization, and depersonalization of state administrative power have also come legal norms and procedures that limit such power according to the principle of democratic accountability. As Gianfranco Poggi has observed, at the same time as the political power of the state has become more extensive in terms of its subject matter and reach, so too have claims for public participation in the exercise of this power widened.[22] This is also to acknowledge the considerable scope for further, more deep-seated democratic oversight. Indeed, it is possible to point to a raft of new ecological discursive designs that have already emerged as partial antidotes to the technocratic dimensions of the administrative state, such as community right-to-know legislation, community environmental monitoring and reporting, third-party litigation rights, environmental and technology impact assessment, statutory policy advisory committees, citizens' juries, consensus conferences, and public environmental inquiries. Each of these initiatives may be understood as attempts to confront both public and private power with its consequences, to widen the range of voices and perspectives in state administration, to expose or prevent problem displacement, and/or to ensure that the sites of economic, social, and political power that create and/or are responsible for ecological risks are made answerable to all those who may suffer the consequences. This is precisely where an ongoing green critical focus on the state can remain productive.

Insofar as any agency of the state (military, police, or environmental protection agencies) is no longer properly accountable to citizens (whether directly and/or via the executive or the parliament), then the democratic state is failing its citizens. Seen in this light, the green critique of the administrative state should be understood not as a critique of the state per se but rather a critique of *illegitimate power.* It is a power that is no longer properly accountable to citizens according to the ideals of liberal democracy. The ultimate challenge for critical political ecologists should not be simply to bring liberal democratic practice into alignment with liberal democratic ideals (although this would be a good start) but to outline a distinctively green set of regulative ideals, and a green democratic constitutional state that is less exclusionary and more public spirited than the liberal democratic state. The concern should not be the mere fact that states exercise power but rather how this power can be made more accountable and hence more legitimate.

4.3 The Ecological Critique of Liberal Democracy

Against the background of the foregoing arguments, the outstanding problem is that liberal democracy is not accountable enough from the perspective of those suffering or concerned about present or future ecological harm. Now some liberals might respond by saying that these worries are misconceived, since the problem is not liberal democracy but rather simply competing human preferences. That is, if, after engaging in lawful means of persuasion and utilizing all available conventional liberal civil and political rights (e.g., holding public meetings, demonstrating, campaigning, bringing legal actions, standing for political office as a green candidate, and voting) an effective majority for wide-ranging sustainability policies cannot be mustered by environmentalists at crucial decision making moments (e.g., general elections, policy making, and law making), then environmental advocates and green parties must simply learn to live with this outcome. Indeed, liberals generally insist that the liberal democratic state cannot, and ought not, guarantee for everyone a good (green) life simply because not everyone shares green values nor interprets or ranks them in the same way. Any attempt to guarantee an ecologically sustainable society would thus be but the latest road to serfdom. Given the fact of competing human preferences, we can only secure sustainability by becoming illiberal, by sacrificing liberalism's openness to individual moral pluralism at the altar of the one true (green) path.

These are now familiar arguments that seek to expose what is believed by liberals to be a failure on the part of radical environmentalists and greens to acknowledge the brute fact of moral pluralism—a problem that is addressed shortly. However, for the moment, what is striking about this standard liberal democratic response is that it insulates liberal democracy itself from any critical ecological scrutiny. Indeed, despite forty years of mounting environmental degradation and persistent public concern and protest, there seems to be a remarkable reluctance among *liberal* political theorists to reflect critically upon the ideals of liberalism or of liberal democracy in the light of the ecological challenge.[23] The question as to whether and/or how far liberalism itself might need to be refashioned to accommodate ecological concerns is a question that has rarely received any lengthy and systematic (as distinct from ad hoc) debate among liberal political theorists, with the exception of a small band of full-blooded libertarians who are committed to "free market environmentalism."[24] The most significant exception to this claim is Marcel Wissenburg's systematic exploration of the relationship between liberalism and environmentalism in *Green Liberalism: The Free and the Green Society*.[25] Wissenburg seeks to discover to what extent *political* liberalism (variously understood as liberal justice or the liberal democratic framework), and philosophical liberalism (its theoretical foundation), can accommodate environmental issues and concerns. Green ideas that can be shown to be incompatible with liberalism are thus discarded. Not surprisingly, Wissenburg concludes that liberal democracy is incompatible with any legal restrictions that seek to alter or dictate people's preferences, whether they are preferences to procreate, drive expensive cars, or otherwise consume. While he concedes that a global Manhattan may not be desirable, he concludes that it is always preferable to an ecologically sustainable society that restricts individual freedoms. At best, the liberal democratic state can facilitate the free flow of information to help citizens exercise informed choice (e.g., by ecolabeling), and it can encourage ecological modernization, of the weak variety, subject only to his innovative restraint principle (discussed below). But liberal democracy cannot control the macro parameters of demand/consumption or population—controls that are basic to any notion of ecological sustainability by ecological economists and radical greens.

Now it is true that unlike liberals, green theorists are more prepared to countenance restrictions on a range of freedoms that are taken for granted, particularly those freedoms relating to investment, production, consumption, mobility, and the use of property.[26] Indeed, green theorists generally maintain that if we are to move toward an ecologically sustainable society and world order, then macro-limits (set by the local community, the state and international community) on economic freedoms are essential. Ecological economists, in particular, have singled out the state as playing a crucial and much more active role in disciplining and channeling market transactions in ways that produce environmentally and socially beneficial outcomes.[27] This is necessary because it would otherwise be more rational for economic actors to privatize gains and socialize costs. From a critical political ecology perspective, however, it is more rational to cultivate ecological citizenship and enable public deliberation over matters of common concern, and if necessary, impose legal restrictions and sanctions to protect public goods by laying down sustainability parameters to ensure that economic activity does not encroach upon ecosystem integrity or biodiversity.

While green political theorists acknowledge the proactive role that green consumers might play in encouraging the greening of investment and production patterns, the important task of laying down sustainability parameters is one that should primarily belong to people acting publicly and democratically as citizens, rather than as consumers. Green consumerism has certainly emerged as an important facet of ecological citizenship, and it has challenged traditional boundaries between the public and the private. However, it cannot substitute for critical dialogue in public fora. That is also why green political theorists are generally skeptical of new ecolibertarian ideas such as "free market environmentalism" that assert that the solution to the tragedy of the commons is the privatization of the commons. Any management regime that seeks to relinquish public control of environmental quality can no longer provide any security against private interests prevailing over the public interest in environmental protection.[28]

Now, in turning the critical spotlight on liberals, critical political ecology does not thereby seek to become illiberal in the sense of pushing for the implementation of the one true green path (as suggested by Wissenburg). Sustainability is an uncertain quest that must be embarked upon in an open-minded, practical, and experimental fashion with the realization that our understanding of ecosystems is not only inadequate but may never be adequate.[29] Nor should critical political ecologists seek to usurp the role of political communities in determining the meaning of, and path toward, sustainability. However, critical political ecologists are concerned to criticise the social and communicative context in which individual and social preferences are formed and exercised. This is not to deny political pluralism but rather to radicalize it.[30] Such radicalization entails exploring the conditions for the flourishing of more plural possibilities, that is, a widening of horizons and options of choice and the creation of more inclusive forms of deliberation. For example, political and economic actors' preferences ought to be challenged in those circumstances when they can be shown to undermine public environmental goods such as the waterways, oceans, atmosphere, and biodiversity, or otherwise restrict the ability of others to exercise their autonomy (e.g., when the health, amenity, or sense of local place of individuals and communities are harmed by the economic choices of agents in other locales, who do not have to live with the consequences of their decisions).

The radicalization of political pluralism also entails questioning the liberal distinctions between public and private, justice and the good—distinctions that emerged in a history of political struggles and power compromises that are now inscribed (and somewhat ossified) in the *liberal* democratic state.[31] Whereas liberal democratic theory enables the "privatization of good" (to borrow Alasdair Macyntyre's formulation),[32] green democratic theory seeks the politicization of the private good as well as the repoliticization of the public good. As Val Plumwood argues, while liberal democracy permits a certain degree of political democracy, it severely restricts citizen participation in precisely those areas that really count in terms of generating ecological problems, such as decisions about consumption, investment, production, and technology.[33] Yet Wissenburg's green liberalism argues that the liberal democratic state ought to avoid regulating these matters because they ought to be left to private enterprise or otherwise be resolved by the exercise of private preferences. These liberal categories also structure the identification, evaluation, and management of ecological risks, including the cultural presumptions and modes of knowing that are brought to bear on risk management. The critical ecological project should be to challenge these liberal "relations of definition" (to adopt Beck's terminology) along with the culture of wardship and the processes of unfair ecological victimisation that they have created. As I argue in chapter 5, this includes reframing the burden and standard of proof, the processes and procedures of risk assessment, and the kinds of knowledge that are relevant to risk assessment.

The problem with Wissenburg's liberal democratic response to ecological problems, then, is that it is far too complacent. It is complacent because it assumes that social structures play no significant role in constituting the individual interests, identities, and preferences of social agents. Rather, all citizens/consumers are considered equally free and unencumbered agents and therefore equally capable of making independent choices, all individuals are fully formed prior to making choices, and all such choices should be accepted at face value. Liberal theorists typically make the rational choice assumption that political preferences are preformed and given prior to economic exchange or political negotiation. They therefore enter the political picture as exogenous variables, and the goal of political actors is to get what they can from the political system via conflict and compromise. The function of liberal democratic politics is to mediate and contain the struggle among self-interested players to pursue their private interests.

In contrast, critical political ecology rests on a relational ontology of the self that recognizes the constitutive effects of social structures—understood in both cultural and material terms. This is not to deny creative agency on the part of individuals. Rather, it is to insist that such agency is always framed by social structures, including established social norms, roles, and identities that fashion the horizons of individual choice. Such an understanding directs attention to the communicative contexts in which environmental policies are formed, whether they are constrained or unconstrained. And it is here that green democratic theorists have argued that policy making in liberal democracies reveals a systematic bias against the protection of public environmental interests and in favor of certain private interests. In other words, the liberal democratic state (and the liberal culture that it both reflects and shapes) is not impartial in the way it prioritizes certain freedoms over others. The upshot is that the liberal democratic state can only guarantee formal rather than substantive freedom for all to determine their own conception of the good.

Environmental protection largely depends on public interest advocacy that is concerned to defend long-term generalizable interests rather than short-term particular interests. Ideally this requires social deliberation and decision making about public goods and interests, rather than political bargaining among self-interested actors in defense of private goods and interests. Critical deliberation is more likely to drive decision making toward the protection of public interests compared to what Charles Lindblom has called "partisan mutual adjustment."[34] The fact that a good deal of policy making in liberal democracies takes the form of partisan mutual adjustment is one good reason why liberal democracies have been unable to deliver more systematic environmental protection.[35] In particular, interest group bargaining is unable to deal with large-scale nonreducible ecological problem, since it tends toward the disaggregation of ecological problems in terms of the particular interests of affected parties. And in the policy bargaining process, it is always difficult to allocate losses, even when the net benefit to society is positive.[36]

Moreover, in the political bargaining over who gets what, when, and how in liberal democracies, political actors who are better resourced, better informed, and strategically located vis-à-vis the centers of policy making invariably have a distinct advantage over socially and economically marginalized groups and classes in the lobbying and bargaining stakes. This advantage is typically reinforced by the cumulative consequences of previous political and social struggles, which disappear into the naturalized background features of the cultural and political landscape to produce the "mobilization of bias."[37] Whereas public deliberation draws out public arguments that must be able to withstand critical questioning from many different vantage points, partisan mutual adjustment narrows the bargaining agenda and favors the more powerful players in the bargaining process. The diffuse notion of the public interest is always at a disadvantage when dealing with a small number of well-organized interest groups with a direct material or financial stake in policy outcomes. Although many new social movements have challenged the rational actor assumptions on which Mancur Olsen's analysis of the logic of collective action is based, the members of new social movements typically represent only a small percentage of the population at any given time.[38] Whereas Olsen would have us believe that the lack of involvement of many citizens in public interest advocacy stems from a rational calculation of the benefits of free riding, there are other, equally plausible explanations such as time and/or income poverty, educational disadvantage, lack of information, and social deprivation, leading to a general disillusionment with the political system as a whole.

Yet there is a more troubling problem at work here in the way in which the political subject is constructed in the processes of political negotiation in liberal democracies. The political struggle in liberal democracies is played out in ways that assume that all political players—from industry associations to public interest advocates—are seeking to advance their own sectional vested interests or private preferences. The political struggle for the limited public favors provided by the state is one that tends to mimic the competitive economic struggle in the marketplace, where it is patently irrational not to behave in self-seeking ways. While ideally public interest advocates may prefer to be engaged in communicative action, they often resort to strategic action in order to win concessions or otherwise make minor advances to further their agenda. Some new social movements take the next step by refusing to play by established protocols and adopting instead disruptive tactics in the public sphere in protest against the distorted communicative context of policy making.

For its part, the liberal democratic state merely offers processes and procedures that seek to channel, mediate, and balance such competing interests (which are made fungible by the assumptions of the rational actor model). Liberal pluralism—in its blindness to gross disparities in social power and communicative competence—asks that all players be equally tolerant of one another and prepared to compromise and live with the outcomes that are negotiated and mediated through the policy-making and law-making procedures of the state. Liberal democracies are thus defended as providing a fair means of reconciling competing preferences, *as if* all preference holders are equally well placed to articulate and assert those preferences. Under these circumstances it is a necessary and desirable part of the bargaining process that environmental concerns are regularly traded off against competing interests. Whereas fundamental liberal freedoms (including freedom of contract) are expressed in the idiom of rights and are entrenched in the formal and/or informal constitutional structures of the liberal democratic state and therefore able to trump competing welfare considerations, environmental concerns are ranked differently. That is, they are considered matters of "the good life" and therefore a matter of competing individual preferences. Unlike liberal civil and political freedoms, environmental considerations are considered nonfundamental and therefore negotiable. This liberal expectation of trade-off and balance tends toward a "short termism" in environmental policy making that is exacerbated by the limited time horizons of political parties and political leaders, who operate within the temporal frame of electoral cycles rather long-range ecological horizons.

Indeed, the utilitarian framework of cost–benefit analysis was inscribed into many of the major innovations in environmental law and administration that took place in the 1970s in most Western countries, notably in environmental, social, and technology impact assessments.[39] However, this framework merely serves to furnish "advice" to decision makers rather than any mandatory directives to the legislature or executive. Moreover the growth of environmental legislation must be understood against an historical background of respect for property rights. This is reflected in the general reluctance by legislatures and courts to impose any restrictions on property rights in the absence of clear proof of harm to others, and the fear of having to pay compensation whenever private property is acquired or "taken" for public purposes. In Anglo-American liberal democracies any harm flowing from the use of property rights has traditionally been regulated by the common law, such as the law of contract and particularly the law of torts, which have placed the onus on those suffering ecological harm or contractual damage to *prove* damage, causation, and dereliction of legal duty. Of course, the spectacular growth of environmental legislation is testimony to the limitations of this traditional method of regulating environmental problems, and property owners face a steadily growing range of legislative qualifications and restrictions to the way they manage and employ their property. For example, the emergence of rules of strict liability in relation to serious risks absolves plaintiffs of the obligation to show negligence (as distinct from causation). Environmental impact assessment procedures are another case in point insofar as they require developers of large-scale projects to show that proposed developments will cause no or negligible harm to the environment. However, many ecological problems arise from the cumulative effect of small-scale activities that undergo no such assessment. While it is possible to track an emerging countervailing discourse of individual or collective environmental rights and entitlements that have increasingly served to qualify property rights, we have yet to see any wide-ranging ecological reconstruction of property rights at the level of principle.[40] Nonetheless, as Gary Varner has argued, the trajectory of development of environmental regulation is such that the day may come when we "treat land as a public resource owned in common and held by individuals only in a stewardship (or trust) capacity."[41] He points out that while the traditional fee simple conception of ownership stresses the unilateral freedom of private property holders to use or dispose of their property as they think fit, the legal positivist conception is of property as a bundle of rights and obligations lends itself to a more radical redefinition. The more environmental regulation restricts property rights and adds to the obligations of the property holder, "the more it chips away at the concept of land as private property."[42]

It is against this background that Wissenburg's exploration of the green potential of liberalism offers a significant innovation that strikes at the heart of classical liberalism's defense of property rights and moves toward something like a stewardship ethic. What he calls "the restraint principle" is partly inspired by John Rawls's savings principle (which is designed to ensure a fair allocation of environmental benefits and burdens among coexisting generations, and indirectly, future generations) and effectively articulates a weak version of the Brundtland Report's formulation of sustainable development. This principle is treated as a distributive principle but not as describing a desirable end-state. Rather, it is defended as a side constraint on the distribution of conditional rights.[43] (Wissenburg uses the term unconditional rights to refer to basic needs, whereas conditional rights refer to further wants.) In essence, the restraint principle provides that rights to scarce goods ought to be, within the limits of necessity/basic survival, restricted in ways that ensure that such goods are available for further use/distribution. That is, ". . . no goods shall be destroyed unless unavoidable and unless they are replaced by perfectly identical goods; if that is physically impossible, they should be replaced by equivalent goods resembling the original as closely as possible; and if that is also impossible, a proper compensation should be provided."[44]

The inverse restraint principle repeats the same formula in relation to the production of pollution (no damaging waste should be produced unless unavoidable, and if unavoidable, nature should be restored and victims compensated).[45] In short, humans may be free to *use* but should not abuse or destroy nature, except where the basic survival of existing generations is at stake. This is indeed a radical proposal that explodes the classical Lockean defense of private property rights along with related notions of finders keepers and *absolute proprietorship* of land and natural resources. In their stead, we are offered a less arbitrary and fairer set of user or custodianship rights that are deeply qualified by responsibilities to coexisting and, indirectly, future generations. The restraint principle thus extends liberal justice to ensure "that individuals get their fairest possible chance to accomplish their plans of life, whatever their plans may be."[46] It is meant to ensure the satisfaction of basic needs while facilitating (without guaranteeing) the satisfaction of further wants. Although defended as an alternative to the precautionary principle (which Wissenburg regards as problematic because it assumes agreement on environment values), it nonetheless serves to place the onus of proof on the environmental exploiter, and not the victims, to show that the principle is satisfied.

Wissenburg's restraint principle provides an example of liberal theory catching up with environmental political practice insofar as it offers a way of conceptualizing the rationale of much contemporary environmental legislation enacted by the modern state. That is, the liberal democratic state—through its legislative, administrative, and judicial agencies—has (to borrow Varner's phrase) gradually chipped away at the unilateral freedom of private property holders to use or dispose of their property as they think fit. Indeed, James Meadowcroft has argued that it is now possible to talk about the emergence of an ecological state, which places ecological considerations at the core of its activities.[47] However, it is arguable that a full-fledged, rather than merely emergent, state of this kind is a postliberal state. Like the welfare state, the ecological state involves an extension of state authority to new areas of social life, provides a response to perceived failures of markets and voluntary action, alters patterns of "normal" economic interaction, represents a continuing adjustment of state activity to new ecological problems, and has complex and contested normative associations.[48] The outlines of this ecological state have emerged within the shell of the liberal democratic state, although it is more of an international than a national creation, its forms of intervention are different, it remains fragmented, and its future is by no means assured.[49] Wissenburg's reconceptualization of the rights of property holders provides another way of conceptualizing the developments in environmental law and policy in recent decades, thereby revealing the antiquated character of the classical liberal defence of private property and its associated understanding of autonomy.

Despite these significant conceptual innovations, Wissenburg's restraint principle rejects any notion of the intrinsic values of nature and is grounded in a thoroughly instrumental posture toward the nonhuman world (with the exception of those animals that bear close resemblance to us). In contrast, an ecological state that upholds the precautionary principle is not only able to regulate the use (by preventing the abuse) of nature but also provide a risk-averse decision-making framework that is able to protect nature for its own sake. This feature, combined with the fact that it is already an emergent international norm, makes it a more promising principle to defend from the point of view of critical political ecology. While the restraint principle makes a radical departure from liberal theory in addressing the fundamental question of the ecological conditions for human freedom, it is still understood as simply restraining rather than constituting freedom. Moreover Wissenburg is careful to emphasize that even the restraint principle cannot provide any ecological guarantees. This is because the satisfaction of even the basic needs of those currently alive may preclude the possibility of saving natural resources and protecting ecosystem integrity for future generations and also because some people may still value other things over a greener present and future. Nonetheless, it is one of those ironies that the most revolutionary ecological reinterpretation of property rights—one that challenges the basic idea of individual and private ownership and effectively converts property holders into ecological trustees with obligations to both present and future generations—has emerged from liberal theory.

4.4 An Immanent Ecological Critique of Liberal Dogmas

The foregoing critique of liberal democracy may be understood as an immanent critique waged on two levels. That is, it seeks to expose the gap between regulative ideals and practices while also calling for a reinterpretation of the fundamental ideal of autonomy and showing how this might be practically realized in terms of an ecological rather than liberal normative ideal. The fundamental problem with the liberal ideal of autonomy is that it rests on an incoherent and undesirable ontology—that of social and biological detachment. Given that ontology precedes ethics (i.e., underlying assumptions about being and reality constrain the field of ethical possibilities), it is necessary to questions these basic liberal assumptions concerning the self before it is possible to rethink what autonomy might mean in a new ecological age.

The fundamental problem with the liberal ontology of the self is that it reduces both human and the nonhuman others to a set of constraints against which, or as instrumental means through which, individual self-realization is to be achieved. The needs and requirements of others are cast as external to those of the lone, self-contained, rational maximizer who, by virtue of what is seen to be a competitive social environment, necessarily enhances his/her autonomy at the expense of others and the environment. As communitarian critics have pointed out, such an understanding of the autonomous self is based on a denial of any noninstrumental dependency on the social world.[50] Critical political ecologists can enrich this critique by showing how liberalism is also based on a denial of any noninstrumental dependency on ecosystems and the biological world in general. Nowhere is this more graphically illustrated than when the liberal self, as economic actor, utilizes property rights in ways that privatize gains and displaces social and ecological costs on to others. As we have seen, the common law of contract and torts has historically provided only a limited repertoire of remedies for such displacement while the public policy process in liberal democracies generates mostly ad hoc, remedial legislation based on a politics of partisan mutual adjustment and therefore continues to favor private interests over public ones. Although Wissenburg's restraint principle provides an innovative ecological challenge to the traditional liberal understanding of property rights, it continues to construct the needs and requirements of others (human and nonhuman) as mere side constraints on individual self-realization.

By emphasizing arm's-length, impersonal contractual obligations over familial, community, and moral bonds, and by seeking to uphold the individual's capacity and right to choose his or her own idea of the good over and above the idea of social deliberation about the common good, apologists for the liberal democratic state deny the fundamentally social character of individual conceptions of the good. As Charles Taylor has argued, community is a structural precondition for human selfhood and moral agency. Individual conceptions of the good can only be acquired and maintained through membership of a language community and culture in which individuals are located. Critical political ecologists can add that ecosystem integrity is a precondition for individual and collective human well-being (in the longer run), and that it can only be properly maintained over time when the human understanding of community is extended to include ecological communities and nonhuman others. Looking after nature becomes not simply a prudent thing to do but also an expression of ecologically embedded selfhood.[51]

This ontological critique of liberalism also challenges the liberal idea of state neutrality in showing how the liberal state reinforces a particular kind of self, with particular kinds of dispositions. To uphold its ultimate ideal (or what Charles Taylor calls the "hypergood") of individual autonomy, liberalism needs not only a liberal political system but also a liberal social matrix that recognizes, protects, and rewards the rational, autonomous self in ways that make it "normal."[52] Thus the liberal democratic state must actively reproduce the social structures that underpin liberal values.

As Bhikhu Parekh has explained in a different context that is nonetheless apposite here: "Although democracy preceded liberalism in Western history, in the modern age liberalism preceded democracy by nearly two centuries and created a world to which the latter had to adjust."[53] In particular, the development of liberal democracy was made compatible with a rapidly developing capitalist form of industrialism, and it was no coincidence that the struggle for political reform came from the rising bourgeoisie. As Marxist-inspired theorists of the state have long emphasized, the social forces that stood to gain the most from capitalism turned out to be the same social forces that were the key players in the process of democratizing the state. Of course, this privileged access to rule making was not to last, as the labor, women's, postcolonial, migrant, and green movements have sought to challenge the rights and privileges of capitalist classes. Nonetheless, the owners and controllers of capital have, for the most part, been more successful than any subsequent social movement in forging the basic constitutional structure and rationality of the liberal democratic state. Moreover, by virtue of their wealth and privilege, such classes continue to have a significant influence in the policy-making process, both nationally and transnationally.[54]

Liberalism not only preceded democracy in the modern age, it also provided its own rationale for the state (to protect the rights of individuals), an account of its formal structure (separation of powers, representative government), an account of the terms on which coercive state power may be exercised (by means of democratic law enacted by the peoples' representatives), and an account of civil society (made up of autonomous individuals). However, once we historicize the particular *liberal* form in which modern democracy has developed, it becomes possible to think about democracy and the state taking on other prefixes, such as ecological. This makes it possible to rethink what role states might play and what form they might take in embodying and giving effect to new social purposes and expanded democratic ideals. Such a rethinking need not require any abandonment of the enduring features of the liberal democratic state, such as the protection of civil and political rights that are essential to ecological citizenship, the election of parliamentary representatives, the separation of powers, the idea that state power should not be absolute or arbitrary but rather limited and exercised according to law, and the idea of toleration and respect for moral pluralism. Rather, critical political ecology should primarily take issue with the limited scope and quality of political representation, participation, and dialogue, and the social and economic structures that constrain political decision making in liberal democracies. The point is to unblock those democratic processes that might subject to critical scrutiny those ideals and practices of autonomy that cannot be generalized for all, including that are not conducive to an ecologically sustainable world. In effect, the quest of critical political ecology may be understood as an attempt to adjust democracy to a world of more complex and intense economic, technological, and ecological interdependence in order to extend the links between environmental protection and social justice. Ecological freedom *for all* can only be realized under a form of governance that enables and enforces ecological responsibility. Ecological democracy is a postliberal rather than antiliberal democracy.

The foregoing critique of liberalism may be seen as seeking to reinterpret rather than reject the fundamental Enlightenment ideal of autonomy. Liberalism's otherwise laudable humanist impulse to expand human autonomy comes to grief in the belief that autonomy can only or best be achieved by mastering the natural world through increasingly sophisticated technologies and the application of instrumental reason. Time and time again—from the splitting of the atom to the building of mega-dams—instrumental rationality has served to imperil rather than expand autonomy for large numbers of people and nonhuman species. As Theodor Adorno and Max Horkheimer famously and prophetically put it, "the fully enlightened Earth radiates disaster triumphant."[55]

A more ecologically informed dialectic of enlightenment therefore requires an engagement with the mutually interdependent ideals of emancipation and critique, or as Tim Hayward has put it, "the twin ideals of mastery and criticism."[56] Emancipation is crucially dependent on critical questioning (of authority, dogma, superstition, or blind faith). Ironically, however, the way in which the basic liberal principle of autonomy has been idealized as self-mastery has served to imperil the development of critical questioning in modern democracies in the new ecological age in ways that have ultimately imperiled autonomy. It is as if liberalism has lost sight of the co-dependence of autonomy and critique by sheltering certain liberal articles of faith from further critical exposure and transformation.

By framing the problem as one of rescuing and reinterpreting the Enlightenment goals of autonomy and critique, it is possible to identify what might be called a mutually informing set of "liberal dogmas" that have for too long been the subject of unthinking faith rather than critical scrutiny by liberals.[57] The most significant of these dogmas are a muscular individualism and an understanding of the self-interested rational actor as natural and eternal; a dualistic conception of humanity and nature that denies human dependency on the biological world and gives rise to the notion of human exemptionalism from, and instrumentalism and chauvinism toward, the natural world; the sanctity of private property rights; the notion that freedom can only be acquired through material plenitude; and overconfidence in the rational mastery of nature through further scientific and technological progress. It is difficult to see how these dogmas would survive critical scrutiny in a genuinely free communication-community in the present ecological age. Indeed, some of these dogmas have already been the subject of scrutiny from within liberal theory.[58]

However, many contemporary liberal philosophers still seem to forget that their liberal forebears forged their political ideals in a bygone world that knew nothing of the horrors of bioaccumulation, threats of nuclear war, Chernobyl and Bhopal, mad cow disease, and global warming. Moreover liberal values were born in an emerging market society that assumed an expanding resource base and a continually rising stock of wealth. Liberalism, along with its great rival Marxism, fully absorbed the Enlightenment idea of progress, assuming that scientific progress and the technological domination of nature would provide plenty for all. These views might have made some sense in the seventeenth and eighteenth centuries, when it seemed more reasonable to suppose that everything about the world was potentially (and soon to be) knowable, available, and rationally controllable. Yet even in John Locke's day, these were somewhat fanciful assumptions, as Locke himself knew. His quaint defense of private property, outlined in the second of his *Two Treatises of Government*, argued that property rights grew out of individuals mixing their labor with nature—which was worthless until appropriated—*as if* there were plenty of unappropriated land for everyone and no one around to object.[59] As this was patently not the case in seventeenth-century England, Locke added the rider that there must be "enough and as good" for others. However, he also made it clear that if all the land was appropriated it could nonetheless be bought with money, which was not equally distributed.[60] Locke's defense of private property thus served to legitimate the unequal appropriation of land.

In contemporary times we have seen that weak ecological modernization still succumbs to the alluring momentum of material progress and the belief in the rational, technological mastery of nature in ways that uphold economic freedoms while ruling out more critical deliberation over the ultimate purpose and character of the modernization process. While weak ecological modernization certainly offers some scope for environmental improvement by promoting a more efficient exploitation of nature, it nonetheless reinforces rather than questions the liberal dogmas. That liberal democracies have so far only managed to work toward weak rather than strong ecological modernization may be taken as testimony of the enduring hold of these liberal dogmas, despite the fact that ecological conditions have changed drastically since the early days of the Industrial Revolution. Policy making in liberal democracies routinely insulates from scrutiny the powerful economic and political interests that stand to gain the most from the perpetuation of these liberal dogmas.

The history of modern grassroots environmental activism and the broader green movement has been, among other things, a history of attempts to address the problems of risk generation and risk displacement by seeking to extend and deepen democracy. In the course of pursuing the cause of environmental protection, these ecology and green movements have sought to improve the quality and to free up the flow and availability of information to affected parties, challenge the entrenched power of technocratic and corporate elites, create more transparency in policy making and administration, and encourage more citizen participation in economic and environmental planning and decision making. In so doing, these movements have brought new issues and concerns to the political agenda, introduced new ways of framing and defining environmental policy problems, and challenged the assumptions and framework of those policy professionals who manage risk assessment.[61] This includes a challenge to the structures of authority that define, assess, and manage risks. In the next chapter I defend a new model of "ecological democracy" that seeks to encompass these concerns by providing a postpositivist, socially and ecologically inclusive model of democratic decision making and risk assessment.

5

From Liberal to Ecological Democracy

5.1 Ecological Democracy: An Ambit Claim

Let us begin with a very simple, but ultimately politically challenging, ambit claim for ecological democracy based on a familiar principle: all those potentially affected by a risk should have some meaningful opportunity to participate or otherwise be represented in the making of the policies or decisions that generate the risk. This formulation is centrally informed by the moral argument that persons and communities should not be subjected to avoidable risk without their free and informed prior consent (I say "avoidable risk" in order to rule out natural environmental disasters involving no or negligible human agency). Now I am not insisting that all those potentially affected by risk must always actually reach a prior consensus about whether or not to proceed with the risk-generating activity, since this is not always practicable (for large populations) or possible (for future generations). Nonetheless, I am suggesting that it is at least intuitively plausible to maintain that if those representatives who do engage in decision making with risk implications for others proceed *as if* all those affected were present, well informed, and capable of raising objections, then this would encourage an orientation that is both (1) risk averse and (2) concerned to avoid the unfair displacement of risk, thereby addressing what can be identified as the "double challenge" of ecological democracy.

Now, at first, there may appear to be nothing new or ecological about this formulation of democracy, as it resonates with those deliberative and cosmopolitan ideals of democracy that seek to incorporate into risk assessment the entire universe of those potentially affected (notably,

Jürgen Habermas's ideal communication community and David Held's cosmopolitan democracy).[1] However, what makes this formulation both new *and* ecological is the accompanying argument that the opportunity to participate or *otherwise be represented in* the making of risk-generating decisions should literally be extended to *all* those potentially affected, regardless of social class, geographic location, nationality, generation, or species. This ecological extension of the familiar idea of a democracy of the affected is intended to be inclusive and ecumenical, incorporating the concerns of environmental justice advocates, risk society sociologists, and ecocentric green theorists. Indeed, ecological democracy may be best understood not so much as a democracy *of* the affected but rather as a democracy *for* the affected, since the class of beings entitled to have their interests considered in democratic deliberation and decision making (whether young children, the infirm, the yet to be born, or nonhuman species) will invariably be wider than the class of actual deliberators and decision makers.[2] As an ideal ecological democracy must necessarily *always* contain this representative dimension, which poses a direct challenge to Habermas's procedural account of normative validity, which runs as follows: "According to the discourse principle, just those norms deserve to be valid that could meet with the approval of those potentially affected, *insofar as* the latter participate in rational discourses."[3] In relation to all those subjects lacking communicative competence, my ecological formulation replaces the words *insofar as* with *as if*. Habermas's procedural account of moral validity rests on the moral principle that ideally persons should not be bound by norms to which they have not given their free and informed consent—a principle that rests on the bedrock Kantian ideal that all individuals ought to be respected as ends in themselves. My ecological account rests on the post-Kantian and postliberal ideal of respect for differently situated others as ends in themselves, and is suitably adjusted to reflect this wider moral constituency. Of course, many nonhuman others are not capable of giving approval or consent to proposed norms; however, proceeding *as if* they were is one mechanism that enables human agents to consider the well-being of nonhuman interests in ways that go beyond their service to humans. Unlike Habermas's formulation, the critical ecological formulation acknowledges the very important role of *representation* in the democratic process. Indeed, this will be the primary basis of my critique of Habermas. And unlike liberalism, my critique also seeks to avoid a purely instrumental posture toward others (whether human or nonhuman) in its extension of the moral principle of "live and let live" to all inhabitants in the wider ecological community, which is understood as an unbounded continuum in space and time.

This reconceptualization of the demos as no longer fixed in terms of people and territory provides a challenge to traditional conceptions of democracy that have presupposed some form of fixed enclosure, in terms of territory and/or people. The ambit claim argues that in relation to the making of any decision entailing potential risk, the relevant moral community must be understood as the affected community or community at risk, tied together not by common passports, nationality, blood line, ethnicity, or religion but by the potential to be harmed by the particular proposal, and not necessarily all in the same way or to the same degree.[4] For example, for a proposal to build a large dam, the community at risk might be all ecological communities in the relevant watershed regardless of the location of state territorial boundaries. For a proposal to build a nuclear reactor, the spatial community at risk might be half a hemisphere, spanning continents and oceans. Temporally this community at risk would extend almost indefinitely into the future, encompassing countless generations. For a proposal to release genetically modified organisms into the environment, the relevant communities at risk might be variable and not contiguous in space or contemporaneous in time. In each case the affected community would typically include both present and future human populations and the ecosystems in which they are embedded. Moreover the boundaries of such communities would rarely be determinate or fixed but instead have more of the character of spatial-temporal zones with nebulous and/or fading edges.

The ambit claim for ecological democracy raises complex moral, epistemological, political, and institutional challenges. It is morally challenging because it loosens the requirement of moral reciprocity that is basic to the Kantian tradition of moral reasoning and conventional notions of citizenship by seeking to extend democratic consideration to a somewhat indeterminate community whose members are not all capable of reciprocal recognition. In this respect it incorporates but goes well beyond liberal cosmopolitan ideals based on respect for the dignity and inherent value of each and every person by recognizing the dignity and inherent value of nonhuman others and the ecological communities upon which both human and nonhuman beings depend.

The ambit claim is epistemologically challenging because it asks those who are able to participate in democratic deliberations to search for meaningful, practical, and parsimonious ways of representing the interests of others who may, in varying degrees, not be fully *knowable* and cannot represent themselves (i.e., future generations and nonhuman others). It seeks to add a new layer to the already vexing question of political representation by adding the concept of political trusteeship: persons and groups within the polity speaking on behalf of the interests of those living outside the polity, for future generation and for nonhuman species.

The ambit claim is politically challenging because it calls for ecological qualifications to the exercise of individual human autonomy (including the exercise of property rights) by repositioning actors responsible for risk-generating activities so that they must literally and/or metaphorically face and answer potential victims, or risk recipients, in an open and critical communicative setting. In short, the demand is that risk generators—whether private property holders or public authorities—must be able to justify their activities in a manner that is either literally or notionally acceptable to potential risk recipients. The failure to provide an acceptable justification to victims and/or their representatives should mean that the ecological risk-generating activity ought not to be undertaken as a matter of environmental justice. This reversal of the burden of proof would have profound consequences on the conduct of both business and government.

The ambit claim for ecological democracy is institutionally challenging because it does not regard the boundaries of the nation-state as necessarily coterminous with the community of morally considerable beings. This poses a direct challenge to the ideas of liberal nationalism and civic republicanism, both of which argue that the proper locus of democratic self-determination should be the national community bounded by culture, sentiment, and the territorial borders of the nation-state. This suggests the need for more flexible democratic procedures that are capable of mapping onto the complex and variable contours of ecological problems and the human and nonhuman communities they affect.

This chapter seeks to defend deliberative democracy as the best means of practically serving this ambit claim for ecological democracy, although it is by no means free from problems. I outline the intuitive green appeal of deliberative democracy after which I will address the moral, epistemological, political and institutional challenges associated with realizing this ideal. An exploration of the international and transnational dimensions of this challenge—including the communitarian argument that genuine democratic deliberation can only take place in a bounded community sharing a common identity—will be held over until chapter 7.

5.2 The Intuitive Green Appeal of Deliberative Democracy

Many green political theorists have turned to deliberative democracy out of dissatisfaction with existing liberal democracy.[5] The primary appeal of deliberative democracy is that it eschews the liberal paradigm of strategic bargaining or power trading among self-interested actors in the marketplace in favor of the paradigm of unconstrained egalitarian deliberation over questions of value and common purpose in the public sphere. That is, the conditions of undistorted and other-regarding communication are defended as more likely to lead to the prudent protection of public goods (e.g., environmental quality) than the distorted and strategic political communication that is characteristic of liberal democracies.[6] Public spirited political deliberation is the process by which we *learn* of our dependence on others (and the environment) and the process by which we learn to recognize and respect differently situated others (including nonhuman others and future generations). It is the activity through which citizens consciously create a common life and a common future together, including the ecosystem health and integrity that literally sustain us all.

Deliberative democracy has a long pedigree, reaching back to Athens, and including the long tradition of civic republicanism as well as more recent innovations in critical theory, such as Jürgen Habermas's discourse ethic.[7] Nonetheless, it is possible to single out three mutually constitutive features that together encapsulate the core ideals and appeal of the deliberative model for those concerned with ecological risks: unconstrained dialogue, inclusiveness, and social learning.

• *Unconstrained dialogue.* The requirement that dialogue be unconstrained or free is a requirement that only justified arguments should be allowed to sway the participants in the dialogue. This requires that participants give reasons for their proposals, reservations or objections to enable the public testing and evaluation of opposing claims. Jürgen Habermas has argued that this requirement for free or undistorted dialogue is anticipated in the very resort to discourse. That is, the presuppositions of communicative reason are that claims can be rationally assessed for their propositional truth, personal sincerity, and normative rightness. The implicit goal of discourse—mutual understanding—can thereby be reached on the basis of "the unforced force of the better argument." Dialogue becomes constrained and distorted to the extent to which participants are swayed by considerations other than rational argument (e.g., by implicit or explicit force, deception, bribery, or the authority and status of the speaker rather than the content of what is said) or when insufficient time is allowed for deliberation over the meaning and consequences of putative facts or proposed norms. The requirement of free dialogue also necessarily encompasses the requirement of publicity. Dialogue is constrained when information is withheld or misinformation is spread. It is also constrained when parties affected by proposed norms are denied an opportunity to participate or be represented in the dialogue.

• *Inclusiveness.* Deliberative democrats typically enlist the requirement of impartiality as an essential requirement of deliberative dialogue, since the point of deliberation is to weed out purely partial or self-interested arguments in favor of arguments that can be defended as acceptable to all. However, in the light of present-day scepticism toward the very possibility of impartial thinking, the notion of inclusiveness, or enlarged thinking, perhaps better describes the other-regarding orientation that is expected of participants (while also avoiding the debate about the possibility of impartiality). Enlarged thinking—or what Hannah Arendt calls "representative thinking"—refers to the imaginative representation to ourselves of the perspectives and situations of other in the course of formulating, defending, or contesting proposed collective norms.[8] This idea of inclusiveness is derived from the more fundamental moral norm of respect for the autonomy of others. Discursive democrats argue that if agreement is to be reached with others, individuals must defend their proposed norms in terms that may be acceptable to others. This is the very mechanism that steers deliberation away from merely selfish arguments toward generalizable ones.

• *Social learning*. The social learning dimension of deliberative democracy flows from the requirement that participants be open and flexible in their thinking, that they enter a public dialogue with a preparedness to have their preferences transformed through reasoned argument. This is, to some extent, a restatement of the requirement of free or unconstrained dialogue, whereby participants are *moved* to change their position by the force of the most appropriately reasoned argument rather than by extraneous considerations. However, this feature also highlights what is typically defended as one of the great strengths of deliberative democracy, that is, its educative and social learning potential. Openness and flexibility on the part of deliberators makes it possible for them to make decisions that are adaptable and self-correcting in face of new circumstances, new information, and new or revised arguments. This is why deliberative democracy is defended as a better candidate than purely aggregative models of democracy (e.g., voting or opinion polling) for enabling reflexive or ecological modernization. Insofar as the latter model merely entails the adding up of individual preferences without any communication or debate between preference holders, it carries less potential for reflexive learning. (Of course, it is also possible to precede aggregation with deliberation; the contrast here is merely between deliberative/communicative and nondeliberative/isolationist preference formation.)

These three features of deliberative democracy—unconstrained dialogue, inclusiveness, and social learning—arguably make deliberative democracy especially suited to dealing with complex and variable ecological problems and concerns. In particular, such a model privileges generalizable interests over private, sectional, or vested interests, thereby making public interest environmental advocacy a virtue rather than an heroic aberration in a world of self-regarding rational actors. It invites reflexivity, self-correction, and the continual public testing of claims. Such critical testing and questioning from the perspective of differently situated others is crucial to arresting and reversing what Habermas once called "the scientization of politics," the process whereby the lay public cedes ever greater areas of system decision making to technocratic (e.g., scientific, professional, corporate, and bureaucratic) elites.[9] The continual critical and public testing of normative claims, including norms embedded in scientific claims, makes it possible to expose and subject to scrutiny the assumptions, interests, and worldviews of technocratic policy professionals, politicians, and corporate leaders. In the field of risk assessment, the deliberative model points toward a more long-range, inclusive, and risk-averse orientation rather than an after-the-fact damage limitation approach. Such a posture necessarily arises when one asks the question: Would *all* those potentially affected by proposed risk-generating practices, rich and poor, citizens and foreigners, now and in the future, consent to such risks if they were fully informed of the potential consequences?

In the terms of the double challenge of ecological democracy, then, deliberative democracy, prima facie, appears promising. Not only is it likely to generate a risk-averse orientation, it is also likely to guard against *unfair* displacement of risks onto innocent third parties. Such an orientation provides a welcome move away from the utilitarian framework of trading-off (which permits the sacrifice of the interests of minorities, those lacking preferences, and the discounted future in favor of present majorities) toward a more inclusive orientation that at least *strives* to find ways of mutually accommodating (rather than trading off) the needs of the present and the future, the human and the nonhuman.

In short, a case can be made that deliberative democracy is especially suited to making collective decisions about long-range, generalizable interests, such as environmental protection and sustainable development. It thus provides a fair process that is likely to move societies toward more reflexive ecological modernization of the kind discussed in chapter 3. Moreover, because it does not confine its moral horizons to the citizens and territory of a particular polity, it may be understood as a transnational form of democracy that is able to cope with fluid boundaries.[10] It also has the capacity to accommodate the complexities and uncertainties associated with ecological problems, include and evaluate both expert scientific and vernacular understandings of ecological problems, and identify and evaluate risks in socially and ecologically inclusive ways. Above all, deliberative democracy may be defended as the best model for reaching mutual understandings about common norms, and the quest to create an ecologically sustainable society is fundamentally a normative concern and only secondarily a technical matter. As John Barry has put it, the concept of sustainability "needs to be understood as a discursively created rather than an authoritatively given product."[11]

Although deliberative democracy emerges as a better candidate for a postpositivist, socially and ecologically inclusive, model of political communication when compared to the bargaining that tends to predominate in existing liberal democracies, it is not without its critics. After all, defending a normative ideal against the imperfect real world practice of liberal democracy begs many questions, not the least of which is how the alternative ideal might be practically realized. Deliberative democracy, particularly Habermas's communicative ideal, primarily addresses the question of the *conditions of free deliberation* but does not entirely grapple with the questions of what beings ought to be included in the circle of moral consideration, how all affected others might be represented (particularly those that are radically different), and how to respond to the problem of political resistance from powerful actors whose position of privilege is likely to be threatened by unconstrained deliberation.[12] Added to these questions are the procedural and institutional challenges associated with trying to realize ecological democracy, particularly in the move from unconstrained argument to the final process of decision making. I will address each of these four challenges in turn, focusing primarily on the question of representing others—a challenge that is especially important in view of disadvantaged minority groups and nonhuman others. The question of future generations will be dealt with more briefly by way of conclusion.

5.3 Representing "Excluded Others": The Moral and Epistemological Challenges

It could be argued that it is only a short step to extend the other-regarding orientation of deliberative democracy to include, at least notionally, nonhuman species as imaginary partners in the dialogue. Following Robert Goodin, we may think of this kind of communication "as a process in which we all come to internalize the interests of each other and indeed of the larger world around us."[13] After all, when the circle of moral considerability is widened to the maximum to include all potentially affected others, *then the very possibility of arbitrarily displacing ecological costs onto innocent human and nonhuman others is foreclosed.* Here it needs to be emphasized that the argument does not assume that we can live in a risk-free society and therefore rule out completely any displacement of ecological costs. Rather, the argument is simply that any such displacement must be *justified* to the satisfaction of the deliberative community (through its representatives), and that the community representatives also be *inclusive*. To fall back on the more conventional idea that *only* human subjects should be recognized as belonging to the kingdom of ends, means that there is no longer any moral objection to displacing ecological costs onto nonhuman nature when there is no obvious cost or backlash on human communities.[14]

As I mentioned in the introduction, this ecological ideal of political communication provides a test for moral validity that goes considerably beyond even Habermas's ideal formulation, which is restricted to communicatively competent subjects. Habermas's procedural account of moral validity rests on the moral principle that ideally persons should not be bound by norms to which they have not given their free and informed consent on the ground that all individuals ought to be respected as ends in themselves. The ecological ideal extends the moral principle to differently situated others as ends in themselves and seeks a kind of notional consent from these communities by means of representative thinking on their behalf, including imaginative role reversal, by those within the dialogue community. The basic argument is that just because all differently situated others may not be capable of providing consent (and this applies to many humans, not just nonhuman others) ought not invalidate the moral claim that, within justifiable and practical limits, all differently situated others (human and nonhuman) ought to be free to unfold in their own distinctive ways and therefore should not be subjected to unjustified policies and decisions that impede such unfolding. Communicative competence is, after all, arbitrary from a moral point of view. As Goodin argues, since the first-best solution of letting all nonhuman others and future generations speak for themselves is impossible, then we either accept the second-best solution of allowing their interests to be represented by others who can speak, or we resign to the unacceptable situation where their interests remain unrepresented.[15]

It is sometimes suggested that it is possible to avoid this second-best solution to the problem of how to represent the interests of nonhuman species by simply focusing on the interests of future generations. That is, if we act fairly and prudently toward future generations by not closing off options, which includes not destroying irreplaceable natural capital, then we will also indirectly serve the interests of nonhuman nature as well.[16] However, such an approach still renders much of nonhuman nature indispensable, and it does nothing to usher in an other-regarding, noninstrumental posture toward the nonhuman world.[17] In any event, my concern here is not to dodge but rather to confront the moral and epistemological challenges associated with the attempt to represent nature for its own sake.

The moral and epistemological challenges associated with this attempt to "speak on behalf of," or otherwise incorporate, the interests of nonhuman others in political deliberation and decision making are in may ways inseparable. Indeed, the philosophical/epistemological debate about how we *know* nature has been central to the controversies surrounding radical green attempts to transcend human chauvinism. It would not be an exaggeration to say that these philosophical questions have continued to stalk those who have sought to defend the very possibility of a nonanthropocentric moral perspective. Clarifying critical political ecology's position on this epistemological debate will also help to clarify critical political ecology's positions on the moral debate.

The project of incorporating nature into the moral community presupposes not only a preparedness on the part of (at least some) humans to take on a trusteeship role but also that these trustees actually know enough about nature to protect it. Yet the concept of nature, and the related idea of the natural, are burdened with multiple and ambiguous meanings, and these meanings are variable across different cultures and over different historical periods.[18] As soon as we historicize the concept of nature, that is, approach it as a complex and shifting social construction rather than an objective reality that is there for all plainly to understand, we raise the question as to whether it is possible to talk about nature as something that is independent of human discourse, and possibly also whether it is something that can be meaningfully represented or liberated.

For example, Steven Vogel has argued that one cannot insist that mute natural entities can be represented in practical discourse by human advocates unless one can provide satisfactory answers to the questions: Who will be the advocates? On what basis? How are they to know nature's interests? He suggests that the problem is different for children or mentally impaired adults, since we can reasonably determine what they would say were they able to speak. This is because we have "some model to fall back on in building our counterfactuals (i.e., normal adult humans); the reasoning involves extrapolating from an abnormal or immature case to the standard one, whereas in the case of animals or other natural entities it involves extrapolating *away* from the normal case."[19] In other words, he suggests it is beyond our ken to imagine what a tree or lake would "say" in response to any proposed norm. Moreover Vogel suggests that the problem cannot be solved by calling in scientific experts, if one accepts that experts cannot be abstracted out from the disciplinary paradigms in which they work; the scientific is always infused with the social and hence the normative and the norms upon which different scientific paradigms may be built can only be determined discursively rather than objectively. The experts themselves cannot be identified other than through discourse. The upshot is that we cannot speak for nature in itself; *we can only speak about the nature we humans have constituted.*

However, none of these arguments need raise a problem for critical political ecology because it eschews naïve realist claims about the world, particularly the idea that there is a direct, unmediated correspondence between human knowledge claims and an objective reality. Rather, knowledge claims about the world, whether scientific or otherwise, are understood as always and unavoidably evaluative, contingent, and filtered through different social frames and social standpoints. Naïve realist understandings of nature are therefore inconsistent with a critical political ecology understanding of the production of knowledge, not the least because they tend to be dismissive of cultural difference (e.g., different taboos about pollution, different cultural and social class orientations toward wildlands and wilderness, and different cosmologies of nature). Naïve realism is also typically blind to the various ways in which the shifting distinction between nature and culture has been used as a form of social discipline and a means of legitimating the exploitation of some humans by others (e.g., the exploitation of women by men, and indigenous peoples by colonial powers). Finally, naïve realism is blind to the way in which scientism denies the validity of local, vernacular forms of knowledge based on experience. In this respect a certain degree of deconstruction of meanings is always required to clear the ground in order to pursue emancipation in ways that do not unwittingly introduce new forms of oppression.

The critical theory dimension of critical political ecology seeks to validate human knowledge claims (whether "facts" or norms, or composite factual/normative claims) by means of critical discourse. Accordingly there would seem to be nothing controversial about the idea that nature cannot be understood independently of human discourse. As Don Marietta has put it: "The only world we can know is the world as it is constituted by consciousness. We have a good substitute for the kind of objectivity many moral philosophers ask for; it is intersubjective verification. Intersubjectivity is as close to objectivity as we can get, and it is close enough."[20]

To the extent that we can reach such intersubjective understandings that transcend particular standpoints (noting that this can be a complicated and hazardous process), we can say we have attained a degree of objective knowledge about the world. Note that such a postpositivist epistemology does not deny the existence of nature as an extra-discursive reality; it simply acknowledges that we do not have any *shared* access to this reality other than through discourse. This necessarily means that we are talking about contingent rather than absolute understanding of objectivity, since intersubjective understandings of "reality" will always be historically and culturally specific, provisional, and potentially always vulnerable to challenge and change.

In contrast, what might be called hyperconstructivist accounts are prone to conflate and collapse the distinction between discourse and reality, while also offering no rational means of resolving conflicting social constructions of reality. From an ethical point of view, hyperconstructivism is necessarily hyperrelativist. It is also politically impotent because it delivers no firm vantage point from which different subjective understandings of nature can be critically evaluated. For critical theory to remain critical, it must necessarily eschew such strong relativism. So I would agree with Vogel that Habermas's basic insight that value is social and communicative must be extended to nature as well (a move Habermas fails to make).[21] However, Vogel takes the argument one step too far in extrapolating from the first claim (that the only nature we know is the nature we have constructed) to the further claim that *there is no nature beyond the nature we have constructed.* Here Vogel is effectively making the hyperconstructionist move in regarding nature as nothing other than what we know through human discourse.

The problem with this move is that it effectively denies "real" extra-discursive nature any independent existence, agency, or creativity on the morally arbitrary ground that it cannot act as a communicative partner. Vogel's analysis recognizes only one form of morally relevant agency in the world—linguistic agency. This leads him to conclude that all of nature is effectively like our built environment, both causally and morally, since we conjure it up through human interaction. This is akin to the problematic Marxian notion of nature as human artifact, or passive material upon which humans leave their mark (albeit understood here in linguistic rather than technological terms). Such an understanding leaves no room for the recognition of any other form of agency in nature, or any relatively independent mode of physical being, or any recognition that nature might be co-partner in biosocial evolution rather than the mere background or stage for the unfolding of human actors.

Yet there is no need to push social constructionism over the edge in this way. Our constructed "nature" is nothing other than our approximate, provisional attempted understanding of so-called real (i.e., extra-discursive) nature. We may think of constructed nature as the ideational map, and real nature as the physical territory. We may discursively constitute nature, but let us not mistake the linguistic map for the manifest physical territory and thereby efface the agency of nonhuman beings and entities. Nature may be our linguistic creation, but it is not entirely our own physical creation. The point is to enable the flourishing of the territory in all its diversity—but we must always necessarily grapple with the fact that we only have shared access to this extra-discursive nature through discursive maps.

If we understand the problem in this way, then there ought to be no necessary *moral* objection to including nature as a subject worthy of consideration in it own right in deliberative dialogue. Rather, the problem is an epistemological one. Moreover the epistemological problem may, when joined with the moral argument, be turned into a positive virtue. That is, the acknowledgment that the only nature we know is a provisional, socially constructed map rather than at best an approximation of the real territory provides the basis of a number of cautionary tales as to how the democratic project might be pursued. Such an argument might run as follows: if we want to respect nature as a relatively autonomous subject yet acknowledge that our understanding of nature is incomplete, culturally filtered, and provisional, *then* we ought to proceed with care, caution and humility rather than with recklessness and arrogance in the way we use and interact with nature. In short, we must acknowledge that our knowledge of nature's limits is itself limited (and contested). Such an acknowledgment is partly encapsulated in the oft-quoted saying that (real) nature may not only be more complex than we presently know but possibly more complex than we will ever know. Practically, these arguments provide support for a risk-averse posture in environmental and technology impact assessment and in environmental policy generally. Morally, it might even mean couching the emancipatory project in largely negative rather than positive terms, since it is often less problematic simply to rule out obviously harmful interventions that it is to engineer what we might believe to be a flourishing nature.

Nonetheless, if it is accepted that communicative competence is arbitrary from a moral point of view, then we ought to accept second-best solutions for realizing these expanded norms of autonomy, since finding an approximate form of representation is better than providing none at all. This means searching for the most efficacious forms of vicarious representation, using the best of our wit, imagination, and current state of learning. In general, expanding the range of environmental information, along with the possibilities of critical interpretation, of human-induced environmental impacts on the widest possible constituency would seem to be the best response to the inevitable limitations associated with any form of political representation.

Practically, this would need to include improving the quality, amount, and free flow of knowledge about ecological problems by means of mandatory state of the environment reporting, along with comprehensive and cumulative (as distinct from merely project specific) regional environmental impact assessments. Extending these familiar mechanisms would go a considerable way toward improving the knowledge base of the general public and environmental policy makers in particular. However, scientific understandings of environmental impact would also need to be placed alongside vernacular understandings of environmental problems based on firsthand field experience by local people (farmers, indigenous peoples) such that the different purposes of knowledge generation for different ecosystems can be laid bare for public scrutiny, testing, and evaluation. However, it is unlikely that the gaps, limitations, controversies, and uncertainties associated with both scientific and vernacular understandings of environmental problems can always be addressed to the satisfaction of all parties, as the question of how to make decisions in the face of value pluralism and scientific complexity and uncertainty will typically arise. Such problems take us directly to the questions of who can speak for whom, who needs to persuade whom, and by what standard? Below, I offer some practical answers to these questions. At this stage I submit that, as a matter of environmental justice, special procedural measures or due process for disadvantaged minorities, nonhuman others, and future generations are necessary to counteract the systematic biases against the interests of this neglected constituency by those existing political actors who might otherwise pursue more short-term, self-regarding economic interests at the expense of these more diffuse and unrepresented interests. As I argued in the previous chapter, it is partly the absence of such environmental due process provisions in liberal democratic states that leads to the routine displacement of ecological costs onto those (whether inside and outside the polity) who lack the means or competence for effective political advocacy. Surrogate forms of advocacy, and decision rules that bring neglected interests into view, provide the best means of protecting the weak from self-serving and exploitative behavior by the strong.

5.4 Representing "Excluded Others": The Political and Institutional Challenges

Those who have attacked the political *feasibility* of deliberative democracy typically argue that it provides too idealized an account of human decision making to be of any practical use in complex, mass societies, where the delegation of political power—including the formal power to deliberate and make laws—is unavoidable. More important, in assuming the absence of concentrations of power in political communication and an other-regarding orientation on the part of deliberators, the deliberative ideal fails to grapple with precisely those things that so often work to thwart fulsome deliberation.

Those who have sought to impugn the political *desirability* of deliberative democracy have typically argued that it is too dispassionate, rationalist, disembodied, masculine, and Western/Eurocentric in its orientation in insisting only on certain modes of rational, critical argument.[22] Moreover they challenge the assumption that ideas are detachable from experience and therefore can be represented in argument by *any* person. Instead, those advocating a "politics of difference" argue that there must always be the presence of persons from minority groups in any deliberative dialogue, and that *what* is to be represented should not overshadow the question of *who* is to do the representing.[23]

Those who attack the *feasibility* of deliberative democracy tend to misunderstand the role of a counterfactual ideal in providing an alternative and critical vantage from which to evaluate and seek to reconstruct political institutions. As a counterfactual ideal, deliberative democracy is necessarily something that is juxtaposed to, and therefore at some distance from, the "real." The point is to highlight what *could* happen *if* certain conditions prevail. As a device for exposing what could be otherwise, the discourse ethic provides a potent critical vantage point from which to unmask unequal power relations and the political actors who sanctify them, identify issues and social groups that are excluded from public dialogue, and sift out genuinely public interests from merely vested private interests. However else one wishes to defend deliberative democracy, I take this "critical vantage point" argument to constitute its unimpeachable core.

Indeed, this same critical vantage point is invoked by critics who seek to impugn the desirability of the deliberative ideal on the grounds that it is too dispassionate, rationalist, and Eurocentric.[24] In pointing to different modes of political communication, such as greeting, rhetoric, storytelling/testimony, and satire, that appear to be excluded from overly rationalistic ideals of deliberative democracy, such criticisms presuppose at least a structurally similar evaluative standpoint to that of deliberative democrats. That is, critics of deliberative democracy effectively join with defenders of deliberative democracy in enlisting the ideal of free and equal human subjects determining their own individual and common destinies in circumstances that are free from explicit or implicit coercion. Without this ideal, there would be no basis upon which to mount such a critique of the status quo. While there is certainly room to argue for a widening of what should count as valid or appropriate political argument or communication, this is still an *immanent* critique that does not in itself impeach the critical normative orientation of deliberative democracy, which is essentially the aspiration to autonomy, understood negatively as not being subjected to arbitrary rule, and positively by having the opportunity to shape the norms that govern collective life.

In any event, deliberative democracy seems well capable of absorbing Young's arguments as well as those who continue to insist that deliberative democracy is impractical. As James Bohman has put it, deliberative democracy cannot ignore different styles of political communication "without threatening social co-operation in deliberation itself."[25] Moreover, if we adopt Dryzek's pithy formulation of deliberation as communication that induces reflection on preferences in a noncoercive fashion, then we leave room for a wide variety of modes of political communication.[26] The ambit claim for ecological democracy effectively employs and extends deliberative democracy in exactly this way—as a regulative ideal of free communication against which we may impugn the legitimacy of the outcomes of real world communication because such communication is unfairly constrained. Here "unfairly constrained" can include insufficiently inclusive in those circumstances where affected parties are not given a voice in the deliberations. This, then, is one (critical) sense in which deliberative democracy is able to serve the ambit claim for ecological democracy.

However, it would be politically unsatisfactory to rest the argument here. In the move from regulative ideals and political critique, on the one hand, to practical institutional reform, on the other hand, many problems still have to be negotiated. These problems arise because, as James Johnson has noted, it is foolhardy to make "heroic assumptions" about the motivations of political actors in democratic deliberation.[27] That is, in a world where power disparities are ever present, it is naïve to expect policy makers always to be so virtuous and patient as to put the public good ahead of their own interests, concerns, and identities and genuinely listen to, and accommodate, all opposing viewpoints in the course of political dialogue and decision making. As Edward Said, in a spirited critique of the discourse ethic, notes: the "scrubbed, disinfected interlocutor is a laboratory creation," which bears no relationship to real political discourse.[28] Moreover the idealizing force of the deliberative model must confront the limitations and practical exigencies of real world political decision making where time, information, and knowledge constraints abound. Clearly, if we are to do justice to the marginal and dispossessed (including those who cannot represent themselves), and if we are to also achieve feasible outcomes, then political procedures and institutions must not be formulated in the philosophical laboratory (where power disparities are absent) but in the real world where power disparities, distortions in communication, and other pressures are ever present.

Moreover, if it is accepted that there is a multiplicity of genres of speech and argument, which may be traced to (among other things) different linguistic and cultural backgrounds, then one might also challenge the normative presupposition of a shared, implicit telos toward mutual understanding in political dialogue, especially in multicultural polities. In such complex and diverse polities, we can expect disagreement to be the rule rather than the exception, and we can also expect that such disagreement will not necessarily always be reasoned or reasonable. Indeed, on many moral, religious and philosophical questions (e.g., the abortion debate), we can expect intractable disagreement.

However, such observations do not render the regulative ideal ineffectual, since without an ideal there would be no normative basis upon which to impugn *any* political communication or decision. Moreover this regulative ideal can still work not only as a *critical* vantage point but also as a *constructive* vantage point, serving as the source of inspiration for ongoing renovations to democratic institutions. As it happens, recent work on deliberative democracy has been increasingly preoccupied with practical concerns about disagreement, feasibility, social complexity, and institutionalisation.[29] Indeed, after an extensive survey of such work, James Bohman has declared that "Tempered with considerations of feasibility, disagreement and empirical limits, deliberative democracy has now 'come of age' as a practical ideal."[30] Many advocates of deliberative democracy have turned their attention away from the counterfactual ideal of deliberation and toward the actual processes of deliberation in an effort to develop a more dynamic understanding of the relationship between ideals and practices. While all deliberative democrats may prize consensus, it is clear that they neither assume nor expect it in every case; instead, they have offered a framework for understanding and dealing with difference and disagreement. For example, Amy Guttman and Dennis Thompson have argued that the fact of persistent disagreement is hardly a reason for abandoning deliberative democracy. Rather, they suggest that it highlights its great virtue, since its procedural requirements (which they identify as reciprocity, publicity and accountability) still make it superior to other methods for resolving political conflicts.[31] It can, for example, better facilitate the search for "an economy of moral disagreement."[32] Similarly John Dryzek has defended "workable agreements," which also resonate with Cass Sunstein's "incompletely theorised agreements," which Sunstein argues "represent a distinctive solution to social pluralism" and "a crucial aspect of the exercise of reason in deliberative democracies."[33] Such agreements are agreements on outcomes and narrow or low-level principles on which people can converge from diverse foundations; they are concerned with particulars, not abstractions. Sunstein also suggests that agreements of this kind are well suited to the need for moral evolution. A turn toward practical, problem solving in the context of cultural pluralism is also the hallmark of the new school of environmental pragmatism.[34]

What seems to emerge from this empirical turn in deliberative democracy is that what I have singled out as the core ideals of the deliberative model—free or unconstrained dialogue, inclusive/enlarged thinking and social learning—sometimes have to be *actively cultivated* or even *imposed* rather than assumed to exist before deliberation, or assumed always to arise in the course of deliberation.[35] When one considers those fora where something approximating genuine deliberation tends to take place (Quakers meetings; university tutorials—at their best; reading groups; well-facilitated public meeting in the town hall; citizens juries; deliberative opinion polls, academic conferences, and consensus conferences), it is because of a preexisting, deep-seated mutual understanding that engenders the necessary mutual respect (the Quaker meetings), a shared culture of critical discourse (the reading group or tutorial), and/or because the forum and its procedures and protocols are carefully *contrived and managed* as a deliberative microcosm to facilitate free dialogue (e.g., the consensus conference). Critical explorations of the group dynamics of conventional juries provides a sobering reminder that small sized groups do not necessarily lead to unconstrained deliberation.[36] This suggests that if deliberative democracy is to be understood as a "school for social learning," then both citizens and their political representatives sometimes need to be actively schooled *in* deliberative democracy before it is likely to take hold and flourish beyond the kinds of enclaves that I have listed. So, in addressing the question of how to "unrig" the anti-ecological biases of liberal democracy, let us accept that the idealized and demanding conditions of deliberative democracy are aspirational, and therefore can only ever be approximated (rather than fully realized) in everyday politics; alternative decision rules other than consensus may need to be applied to foreclose what might otherwise be interminable debate or to respect cultural difference, and, as Young argues, cultural and social differences should be considered a resource for public reason, rather than as divisions that public reason must somehow transcend.[37]

Now when we turn to these practical challenges we find that the problems raised by Vogel in relation to "speaking for nature" merely represent an extreme version of a more general and enduring problem concerning the unavoidable epistemological and motivational hazards associated with *all* forms of political representation.[38] That is, there are many reasons why political representatives may find it difficult or impossible to understand or imagine the perspectives of *all* differently situated others in order to formulate norms that may be acceptable to those others. This may arise because of lack of personal experience of the other, lack of information, or misinformation, or scientific uncertainty. Or it may arise because representatives lack the necessary motivation to treat the lifeworld and interests of differently situated others on an equal par with their own. More generally, as feminist difference theorists have pointed out, *all* political arguments, however well intended, cannot be entirely detached from the experience, cultural and class background, and material interests of their proponents.

These epistemological and motivational deficits associated with political engagement and political representation cannot be eliminated from political life. However, they can be minimized and/or held in check by a range of institutional devices that make it difficult for parties to act corruptly, deviously, or even just self-interestedly while also encouraging long-range, inclusive deliberation. Without offering an exhaustive response to the challenging question of institutional reform, I will suggest a number of such devices (some familiar, some less familiar) that might help to bring into fuller view the community at risk. Now I have already noted that it is neither possible nor practicable for all affected parties literally to deliberate together *en masse*. Indeed, ecological democracy must necessarily contain a representative element if it is to function as a democracy *for* the affected, including future generations and nonhuman species.[39] Accordingly the question of political representation emerges as a crucial issue in both the theory and practice of ecological democracy.

The first and most significant step is to support mechanisms that ensure that political representation is as *diverse* as possible. In short, deliberative democracy must be representative in a double, reflexive sense. It must encourage enlarged thinking, and it must also provide for enlarged, as in diverse, representation on the understanding that it is dangerous always to "trust" in the political imagination of the chosen or privileged few (Burkean, Madisonian, green, or otherwise). While it is impossible to orchestrate a meeting of the entire community at risk, we can at least devise forms of political representation (along with appropriate procedures and decision rules) that serve to widen and deepen the horizons of those who are actually engaged in the making of risk-generating decisions. In particular, risk-generating and risk-displacing decisions are less likely to survive policy-making communities and legislative chambers that are inclusive in terms of class, gender, race, region, and so on, and especially so when the deliberators are *obliged* to consider the effects of their decisions on social and ecological communities both within *and beyond* the formal demos. Such procedures would, in effect, serve to redraw the boundaries of the demos to accommodate the relevant affected community in every potentially risk generating decision. (Such procedures also offer an alternative to the multiplication of regional and international governance structures advocated by cosmopolitan democrats that introduce their own democratic deficits.)

Diverse representation guards against self-interested collusion and also facilitates enlarged thinking by minimizing the problem of a narrow band of elites "second-guessing" (benignly or otherwise) the concerns and interests of differently situated others, especially minority groups. In the language of Anne Phillips, the "politics of ideas" must be supplemented with a "politics of presence."[40] This argument is also broadly consistent with Iris Marion Young's neo-Habermasian conception of communicative democracy, which criticizes both the liberal and civic republican ideal of impartiality and relies instead on group representation as a strategy of displacement in relation to entrenched ways of framing and responding to political problems by political elites. Diverse representation provides one means of confronting, displacing and ultimately stretching the political imagination of representatives, thereby going some way toward correcting the exclusionary implications of the knowledge and motivational deficits associated with all forms political representation. From the point of view of environmental justice advocates, ensuring the presence of racial minorities or disadvantaged groups in legislative assemblies and environmental policy making communities (e.g., via balanced tickets and proportional representation electoral systems) will go some way toward preventing the unfair displacement of ecological and social costs onto those minority communities. The adoption of multimember electoral systems would also increase the likelihood of green parties gaining a formal presence in parliaments.

Yet diverse cultural representation in policy communities and legislative chambers must also be supplemented with specialised environmental advocacy. The presence of green parties obviously goes some way toward providing more systematic representation of environmental concerns, but green parties are typically poorly resourced and still politically marginal when compared to mainstream parties backed by wealthy and vested producer interests (i.e., capital, and to a diminishing extent, organized labor). Dennis Thompson has suggested the establishment of forums in which representatives could speak for the ordinary citizens of foreign states while also considering the views of international organizations, possibly formalized by a Tribune for Noncitizens.[41] The point of such an innovation is to correct the bias of commercial and government-to-government negotiations. In a similar vein, Andrew Dobson has defended the provocative idea of proxy representation of both nonhuman animals and future generations in representative assemblies by deputies elected from the environmental sustainability lobby.[42]

There are many other ways in which environmental advocacy might be institutionalized in the policy-making process to complement and challenge (as distinct from replace) the advocacy of environmental nongovernment organizations (NGOs) and grassroots community groups in civil society. At the local and national levels, the establishment and proper resourcing of an independent environmental defenders office, staffed by a multidisciplinary team and charged with the responsibility of environmental monitoring, political advocacy, and legal representation would go some way toward ensuring that more systematic attention is directed toward the nonhuman constituency.[43]

However, while expanding the range of voices (including proxies or trustees) in democratic deliberation will go someway toward redressing the power trading and short-termism of liberal democracies, it does not provide a complete answer to the question of how to make actual decisions in the face of value pluralism, conflict, and scientific complexity and uncertainty.

In cases of scientific uncertainty and conflict between environment and development interests, the democratic state cannot be neutral. It can either support the status quo, which favors property holders and risk generators, or create new rights and new presumptions that turn the tables in favor of environmental victims. The requirements of environmental justice that are embedded in the ambit claim for ecological democracy demand rights and decision rules that positively favor the disadvantaged and communicatively incompetent over well-resourced and strategically oriented economic actors in cases of uncertainty and political intractability.

One such mechanism for shifting the presumption in favor of potential environmental victims is the precautionary principle. The Rio Declaration formulation of the precautionary principle provides that "Where there are threats of serious or irreversible damage, lack of full scientific certainty shall not be used as a reason for postponing cost-effective measures to prevent environmental degradation" (principle 15). Adding the words "to present and future human and nonhuman communities" after the words "irreversible damage" would head off narrow, anthropocentric interpretations of this decision rule, which provides a presumption *against* decisions carrying serious or irreversible environmental risks (e.g., species extinction, climate change, nuclear fallout, and so-called genetic pollution from the release of genetically modified organisms into the environment). The decision rule also serves as an evidentiary rule in placing the onus of proof on the proponent to prove the *absence* of such risks for human and nonhuman communities, now and in the future.

Of course, the precautionary principle would need to be interpreted and applied discursively in particular cases. However, participants in the dialogue should not be free to ignore it. One way of ensuring this is to constitutionally entrench the precautionary principle in the same way that basic democratic rights are constitutionally entrenched. Such entrenchment would not place the precautionary principle beyond the reach of democratic debate, since the appropriateness of its application to particular circumstances would always need to be debated on a case by case basis. In any event, the justification for entrenchment is itself a democratic one: to ensure that the interests of those at risk who cannot be present are nonetheless systematically considered by those who are present. Mandating such consideration is not the same as mandating particular outcomes.

Of course, there may be other legislative rather than constitutional ways of making this procedural rule a requirement of public decision making. Either way, the precautionary principle provides a highly effective and parsimonious means of forcing more systematic consideration of potential environmental impacts on differently situated others, including impacts on the interests of future generations and nonhuman species.[44] The precautionary principle has already been widely adopted in international, regional, and domestic sustainable development strategies and policies, and increasingly incorporated into legislation, although there is further scope to specify the meaning and application of the principle in more detail.[45]

The case for systematic enlistment of the precautionary approach (whether through constitutional entrenchment or others means) may be justified on grounds of fairness: that special constitutional protection is required precisely because there are no other direct and formal mechanisms to ensure the representation of the nonhuman world, future generations, and "noncitizens" living beyond the polity, even though they may be adversely affected by decisions made within the polity. The entrenchment serves to prevent or minimize problem-displacement and is therefore anticipatory rather than merely compensatory.[46] No single decision rule is likely to do more to protect environmental victims.

The constitutional entrenchment of the precautionary principle may also be linked to the recognition of a human right to environmental protection. Support for such a right is gaining ground, and the UN Special Rapporteur on Human Rights and the Environment Fatma Zohra Ksentini has proposed the adoption of a set of Principles on Human Rights and the Environment.[47] Moreover more than 70 countries already have environmental constitutional provisions, although not all these provisions are specified in terms of enforceable rights.[48] In his detailed examination of the arguments for constitutional environmental rights, Tim Hayward has argued that the case for *procedural* environmental rights is "all but unanswerable" while the moral case for substantive environmental rights is unimpeachable insofar as a basic environmental minimum is a precondition for democratic decision making.[49] Worries that such rights might confer too much power on the judiciary or constrain or compromise future democratic decision making can be answered primarily by pointing out that such rights are designed to enhance rather than foreclose democratic debate. As Ronald Klipsch explains, the point is to create an environmental due process that minimizes judicial involvement and broadens democratic processes.[50] Such rights might include a right to environmental information (and a corresponding duty on the part of the state to provide regular state of the environment reports), the right to be informed of risk-generating proposals, third-party litigation rights, a right to participate in environmental impact assessment processes, and the right to environmental remedies when harm is suffered.[51] At the same time these legal gateways make it possible to redress environmental injustices.[52] That is, constitutional environmental rights would also underpin changes in legal presumptions (particularly regarding property rights and the onus and burden of proof) in ways that enable a new and fairer balance to be established between concentrated and well organized economic interests and emergent, diffuse public environmental interests. As discussed in chapter 7, these rights can, under appropriate circumstances, also be reciprocated by multilateral arrangements between states and thereby become transnationalized.

The institutional innovations sketched above are illustrative only, and doubtless there are other ways of institutionalizing environmental justice. Inclusive deliberation also demands more inclusive forms of representation and new, ecologically sensitive procedures and decision rules if ecological justice is to be done. This is not a case of rigging the system in favor of the environment, as some liberal democrats might wish to argue. Rather, the suggested procedures and decision rules—inspired by the ambit claim of ecological democracy—are intended to *redress* the unequal power relations that routinely thwart fulsome risk assessment on behalf of the community at risk. The point is to improve the conditions and inclusiveness of dialogue by redressing major power imbalances in political communication and representation.

Now it might be argued that the ecological democracy that I have defended here is not really a radical *departure* from liberal democracy, merely a radical *extension* of it. As we have seen, the principle of considering all affected interests is a familiar one. My response to this argument is to reiterate that ecological democracy is a post-liberal democracy, not an anti-liberal democracy, and it is arrived at by means of an immanent critique of existing liberal democratic regulative ideals and practices in the same way that social democracy (and democratic socialism) emerged out of a critique of classical liberalism. Accordingly, one can always point to continuities and discontinuities. Although the institutional renovations that I have suggested would initially appear as extensions to the existing edifice of the liberal democratic state, they may, over time, come to redefine the rationale and purpose of the state. And although these renovations are mostly procedural, it should be clear that no model of democracy—whether liberal or ecological—is merely or entirely procedural. All models of democracy are derived from more fundamental norms that, in turn, reflect and shape the character and boundaries of the moral and political community that the designers seek to cultivate. At the end of the day, the relative merits of liberal versus ecological democracy ought to be judged not simply in terms of their procedures but also in terms of the ultimate values they seek to serve and uphold.

6

The Greening of the Democratic State

6.1 From Ecological Democracy to the Green Democratic State

In the previous chapter ecological democracy was defended as being far more conducive than liberal democracy to furthering environmental protection and environmental justice. I also suggested some practical procedures and decision rules, such as the precautionary principle, that could be incorporated into the formal political and legal decision-making procedures of the state.

Yet it would be misleading to think of the green democratic state as merely the liberal democratic state with a few of these ecological democratic innovations bolted on around the edges (although this may well be the way in which the green democratic states emerge). Rather, these procedural innovations should be understood as intimations of a new type of post-liberal state, based on a new rationality and normative purpose that builds upon but also challenges some of the basic presuppositions and values of liberalism and the liberal democratic state. Indeed, if any deeper transformation of liberal democracies is to take place, it is essential that such procedural innovations be pursued in the context of a more fundamental debate about the rationale and ideals of the democratic state in an age of ecological crisis.

The purpose of this chapter is to locate the democratic innovations defended in chapter 5 in the context of recent critical theories of the state, civil society, and the public sphere. The liberal democratic constitution presupposes and seeks to maintain a liberal notion of public reason that recognizes, protects, and rewards rational, autonomous, and freely choosing individuals in both the economic and political realms. A green democratic constitution would likewise require the flourishing of its own kind of public reason—in this case, critical ecological reason—that recognizes, protects, and rewards ecologically responsible social, economic and political interactions among individuals, firms and communities. The point of drawing attention to these parallels is to show that the green democratic state is no more or less normatively loaded than the liberal democratic state in that both seek to uphold particular rules that promote certain ethical patterns over others. Accordingly the real contention between liberals and greens is not whether one state is neutral while the other is not but rather which set of rules (and the patterns they tend to promote) might be defended as more desirable and legitimate in the context of highly pluralized societies confronting complex ecological problems.

It is a central contention of this inquiry that the green democratic state is more desirable and likely to be *more legitimate* by virtue of the ways in which it seeks to extend and deepen democracy to hitherto excluded others. The aspiration is to transcend the (uncritical) ethical subjectivism of liberalism by offering a more critical, intersubjective assessment of agents' preferences without stifling cultural and moral diversity. However, the green democratic state cannot be relied upon alone to uphold these processes and in any event must always be understood as part of a broader, state-society complex. States and societies are connected by the public sphere, comprising those communication networks or social spaces in which public opinions are produced. One of the aims of green constitutional design should be to facilitate a robust "green public sphere" by providing fulsome environmental information and the mechanisms for contestation, participation, and access to environmental justice—especially from those groups that have hitherto been excluded from, or under-represented in, policy-making and legislative processes. Such mechanisms are not only ends in themselves but also means to enhance the reflexive learning potential of both the state and civil society.

Whereas the freedoms enjoyed by citizens in the classical liberal democratic state have been traditionally understood to be "pre-political" freedoms that existed prior to, and were subsequently upheld by, the constitutional framework of liberal democracy, the freedoms enjoyed by citizens in the green democratic state are always necessarily *constituted by* the constitutional framework and democratic public law that is enacted under the aegis of the green democratic constitution. Classical liberalism permits but does not demand critical public discourses to uphold its conception of freedom as noninterference. In contrast, a critical ecological conception of freedom demands such a discourse and in this respect has more in common with the civic republican conception of freedom as active participation in the life of the polity.

However, the civic republican tradition has long been stalked by the problem of how to convert the "will of all" (each individual will) into the "general will" (or common good) in noncoercive ways. (Rousseau, in particular, failed to grapple with the question of representation, and he sidestepped the difficult question of majority-minority relations.) The existence of irreducible moral and cultural pluralism, not to mention structurally antagonistic material/economic interests, within civil society poses a major challenge to the quest to uphold a green democratic state along traditional civic republican lines. Such differences call into question the very idea of all citizens leaving behind their private or particular interests in the search for the common good while also highlighting the diversity of potential conceptions of the common good. Critical political ecology's challenge is to develop a constitutional framework that is able to foster political tolerance of diversity but without lapsing into cultural relativism or a defense of political bargaining and partisan mutual adjustment or uncritical preference aggregation (which is the characteristic mode of dealing with moral pluralism and antagonistic social forces in liberal democracies). This entails navigating a path that avoids the homogenizing and potentially oppressive tendencies of civic republicanism as well as the socially fragmenting tendencies of liberal democracy.

This chapter will seek to work toward such a normative theory of the green democratic state by means of a critical dialogue with the most influential rival theory to both liberalism and republicanism, notably, the discourse theory of law, democracy, and the state offered by critical theory's most influential contemporary scholar—Jürgen Habermas. As it happens, Habermas situates his theory of discursive democracy, the state, and law *between* liberalism and republicanism in an effort to combine the virtues and minimize the weaknesses of both. An immanent critique of Habermas from the perspective of critical political ecology will thus perform a double critical duty. The focus of this chapter will be the domestic face of the green democratic state; Habermas's position vis-à-vis the external face of the green democratic state is pursued in the next chapter.

Although I seek to show that Habermas's critical theory of the state is ultimately not critical enough, his account of the relationship between the state, civil society, and the public sphere nonetheless provide a fertile basis from which to reconstruct the green democratic state. In particular, it highlights the centrality of the emerging green public sphere to the political project of greening the state.

6.2 The State, Civil Society, and the Public Sphere

Whereas liberal theories of justice set out to clarify and limit the role of the state, and articulate the circumstances when state coercion is justified, Habermas has (traditionally at least) been much less explicit about the role of the state. One reason for this is that he originally developed his account of the discourse ethic to explain communication in the public sphere, which he understood to be located *outside* the institutions of the state and which was to serve as the unconstrained forum for discussion of public norms.

Beginning with his pioneering work on *The Structural Transformation of the Public Sphere*, Habermas has long been concerned to recover and reinterpret for the present age emancipatory ideals that were born in eighteenth century Europe.[1] For Habermas, emancipation could be achieved through rational-critical debate about matters of common concern by free and equal individuals in the public sphere. The independence and critical publicity of the early bourgeois public sphere—an arena in which private people came together as a public—thus held out the promise of putting an end to domination.[2] In this early formulation, the public sphere was not coterminous with the state apparatus, since it consisted of all those who might join in a discussion of political issues raised by the administration of the state. Moreover it was the very independence of the early bourgeois public sphere that enabled critical publicity.

However, to maintain and develop this potential, the bourgeois public sphere had to become inclusive rather than exclusive by admitting other social classes and groups that had traditionally been excluded from the coffee houses and salons attended by the literate middle classes of eighteenth-century Europe. Yet the more the public sphere expanded and become heterogeneous and open-ended, with no shared class consciousness or belief system, the more difficult it became for it to function as a critical-rational forum.[3] Moreover the combined effect of the rise of mass communication, the culture industry, mass advertising, and mass political parties (with their slick public relations "machines") has meant that the promise of critical publicity held out by the early bourgeois public sphere has been degraded into an ideology, manifested as a set of public opinion molding practices that are routinely staged by political parties and political elites for the purposes of acclamation.[4] As Habermas puts it, "the 'suppliers' [of public opinion] display a showy pomp before customers ready to follow."[5]

Nonetheless, in both his early and subsequent explorations of the public sphere, Habermas has continued to emphasise at least the *potential* of the public sphere to act as a source of critical reason, noting that "public opinion can be manipulated but neither publicly bought nor blackmailed."[6] Although not originally intended as a model for democracy writ large, Habermas's early ideas about the public sphere have also served as the theoretical inspiration for a critical conception of deliberative democracy or "discursive democracy" (as Habermas now tends to call it). To the extent to which the state has subsequently been brought into this picture, its proper role is to foster the conversational spaces that enable legitimate discourse to occur within the state *and* civil society.

Whereas in *Structural Transformation* Habermas's solution to the withering of the public sphere in mass democracies was to proceed with "the long march through the institutions" (including political parties, parastatals, and the bureaucracy) by excluded classes, in his more recent work he has emphasized instead the ways in which the public sphere is structurally distinct from other action coordinating systems such as the market or administrative rule because the public sphere is rooted in the "lifeworld" through the associational networks of civil society.[7]

However, the processes of modernization that have enabled systems of money and (bureaucratic) power to taken over from linguistically mediated expectations in coordinating social action has led to a lopsided rationalization that has *depoliticized* the domination of the lifeworld by capitalism and the administrative state.[8] Thus the separation between the powerful actors on the main stage and the passive spectators in the public gallery has persisted. Nonetheless, Habermas continues to point to those more spontaneous, critical, and grassroots initiatives, associations and movements within civil society as being able to both utilize and continually radicalize political communication structures.[9] Herein lies the means by which the "sleeping gallery" or "the public sphere at rest" may be awakened in ways that carry the potential to "shift the entire system's mode of problem solving."[10]

In *Between Facts and Norms*, Habermas provides a systematic extension of the discourse principle beyond the loose and informal processes of opinion-formation in the public sphere to include formal democratic will formation or law making by the legislature, and legal adjudication by the judiciary. Here we find Habermas's most systematic attempt to "bring in the state" (and its associated legal system), and conceptually relate it to civil society and the public sphere. Indeed, he now maintains that the discourse principle can only be upheld and realized *within* the framework of a democratic legal system, which upholds those rights that are necessary for free discourse to occur. In this new "constitutional turn," Habermas has argued that basic constitutional rights do not stand prior to democracy, as the first principles of justice, in the way that John Rawls had argued. Rather, basic rights and democracy are understood as presupposing each other and are therefore understood to be mutually constitutive.[11]

While Habermas's account of democracy and the state appears not dissimilar to that of liberal democracy, he is nonetheless at pains to situate discursive democracy *between* liberal and republican/communitarian models of democracy. Yet this repositioning is based on somewhat caricatured understandings of both liberalism and republicanism. By "liberal" Habermas has in mind the Lockean tradition that sees the state as primarily the guardian of an economic (read capitalist) society. Yet even he concedes that there are many egalitarian liberals, such as John

Rawls and Ronald Dworkin, who would reject this tradition (and it is sometimes difficult to distinguish Habermas's own theory from Rawls's political liberalism).[12] By republicanism, Habermas has in mind the defense of a substantive ethical community that is institutionalized in the state. Again, this nostalgic rendering of republicanism has been considerably reworked by contemporary republicans (e.g., Philip Petitt), whose defense of contestation and deliberation resonates strongly with Habermas's discursive democracy.[13] Nonetheless, the outdated accounts of liberalism and republicanism set up by Habermas serve as a convenient foil for his defense of a procedural understanding of democracy.

According to Habermas, the thin (Lockean) liberal view expects democracy to take the form of strategic bargaining among competing interests. The democratic constitutional arrangement provides rules that enable compromises to be reached, which are ultimately justified in terms of upholding basic liberal rights. Under this liberal arrangement the gap between the state and society cannot be eliminated in highly pluralized societies, only bridged by representative democracy and the rule of law. The point of the liberal constitutional arrangement is merely to ensure that the power exercised by the peoples' representatives is not abused, and that fair compromises are reached that reconcile competing preferences. Yet Habermas argues that in rightly forgoing the unrealistic assumption of a citizenry capable of collective action, this restricted liberal understanding of politics aims too low in merely seeking to uphold the private expectations, interests, and plans (or "good") of its citizens.[14] The idea of a common good or generalizable interests falls away and instead we have the highly fragmented will of all, which has no political (or ecological) rationality.

Yet for Habermas the republican/communitarian alternative is also problematic because it suffers from what he calls "ethical overload" in its unrealistic expectation of a shared cultural background.[15] Moreover Habermas has criticized the republican concern to find ways in which "popular sovereignty" or direct, decentralized self-governance might be enacted, as if the gap between the state and society could be closed off in favor of a society-centered account of democratic governance.[16] By insisting that sovereignty should remain in "the people," sovereignty could not be delegated and "the people" could not have others represent them.

The challenge for critical theory, then, is to uphold an understanding of democratic politics and the state that avoids the liberal compromise of merely balancing or trading off individual preferences while also avoiding the utopian (and potentially tyrannical) republican alternative of transcending all differences in an effort to arrive at the common good. For Habermas, the way to avoid these two unpalatable alternatives is to institutionalize and uphold the *procedures* that ensure democratic deliberation. Such a proceduralist conception of discursive democracy is defended as avoiding the mere clash of sectional interests while also allowing partial interests to be expressed. Such a conception is also proclaimed to acknowledge cultural diversity and moral pluralism while also providing the basis for actively constructing, by means of discursive dialogue, the necessary social solidarity for common problems to be dealt with in common ways. As Habermas explains, "Discourse theory takes elements from both sides [from liberalism and republicanism] and integrates these in the concept of an ideal procedure for deliberation and decision making."[17] Thus Habermas seeks to uphold the republican defense of public or collective reason over individual rational choice strategies while also pragmatically accepting representative democracy and delegated power, guided by the principles of the constitutional state. Discursive democracy, he argues, combines both such that "the 'self' of the self-organising legal community disappears in the subjectless forms of communication that regulate the flow of deliberations in such a way that their fallible results enjoy the presumption of rationality."[18] In Habermas's reconstruction of the democratic constitutional state, popular sovereignty is reinterpreted intersubjectively (it "retreats into democratic procedures").

The success of this de-centered, and legally secured, account of discursive democracy depends on the proper institutionalization and acculturation of the procedures and conditions of free communication in both the state and a de-centered, highly pluralized civil society. These procedures and conditions are also understood as providing the basis for the development of social solidarity among strangers in relatively autonomous public spheres in civil society and the more formal deliberative arenas of the state.[19] More recently, in his exploration of the various multicultural strains acting upon the old idea of the nation, Habermas has defended the idea of "constitutional patriotism" by which he means a shared commitment to democratic procedures based on the principles of mutual recognition and political equality.[20] Constitutional patriotism is defended as providing a more abstract, more cosmopolitan, and more inclusive foundation than ethnicity upon which to ground social solidarity.

While the role of the democratic constitutional state is to uphold the rights and procedures for democratic opinion- and will-formation, the lifeblood of the democratic polity is derived from, and continues to reside in, civil society and the public sphere.[21] Although the public sphere as a network has taken on increasingly abstract and virtual forms with the rise of the new communications technologies such as the internet it nonetheless remains for Habermas nonspecialized in the sense that "it is tailored to the general comprehensibility of everyday communicative practice."[22] Social actors in civil society utilize these networks not only to publicize problems but also to thematize and dramatise them in ways that might be taken up by political parties and legislators. Drawing on the work of Jean Cohen and Andrew Arato, Habermas defends the idea of new social movements conducting a "dual politics" that is directed toward the state, on the one hand, while also maintaining critical distance from the state, and consolidating new collective identities in civil society, on the other hand.[23] He also suggests that this can be achieved or aided by the radicalization of existing rights, although he does not develop the point.[24]

Now Habermas acknowledges that the political "influence" of civil society, via the public sphere, is only converted into political power when it actually affects the beliefs of those who are constitutionally authorized by the political system to make binding decisions. While this might appear to give state elites the upper hand in the "constitutional deal," he also insists that the political system in democratic societies must remain sensitive to the public sphere if it is to retain legitimacy (primarily, but not exclusively, through political parties and elections).[25] In this sense, at least, the state depends on the approval of civil society, voiced through the public sphere, understood as "an *intermediary structure* between the political system, on the one hand, and the private sectors of the lifeworld and functional systems, on the other."[26] Moreover, while the public sphere is fluid, amorphous, and unbounded, democratic law necessarily always gives expression to the particular wills of the representative members of a particular legal community. That is, once the discourse ethic is institutionalized in the state, deliberative politics becomes unavoidably anchored in, and restricted to, a *particular* existential legal community and is therefore necessarily delimited in space and time "with specific forms of life and traditions."[27] The modern democratic constitutional state—a relatively recent development in human history—must establish legitimate procedures of law making as well as "the *language* in which a community can understand itself as a voluntary association of free and equal consociates *under law*."[28]

The negotiation of legal norms may also accommodate different kinds of discourse, and different kinds of argument and justification. Indeed, Habermas has now developed the discourse principle in quite flexible terms to accommodate moral-political, ethical-existential, *and* pragmatic modes of argumentation, which correspond with the just, the good, and the purposive (or instrumental).[29] Each provides a distinctive mode of reasoning, and a distinctive orientation to others, in response to practical questions. Whereas moral deliberation requires a perspective freed from personal, egocentric, or ethnocentric considerations, ethical deliberation is oriented to the telos of particular lives and communities. Pragmatic arguments, in contrast, take as their starting point the particular preferences and goals of particular agents. The preferences and goals of other agents merely serve as limiting conditions, and each agent assumes that others will be acting according to their own interests.

Habermas's point in making these analytical distinctions is to clarify the *formal* aspects under which legal norms are different from moral norms. That is, the practical test for what is *politically legitimate* is by no means as stringent as the ideal test for *moral validity*. Political legitimacy demands only that norms emerge from *fair procedures* rather than shared lifeworlds. While moral norms are ideally based on a rationally motivated *consensus*, legal norms need to be based on a rationally motivated *agreement*.[30] Legal norms specify what parties must (or must not) do, creating enforceable rights and responsibilities. This more historically concrete character of legal norms (when compared to more abstract moral norms) affects the content of the law, the meaning of legal validity, and the mode of legislation.[31] These requirements flow not only from the need for legal certainty but also from the need to ensure the equality of legal subjects before the law.

Legal validity within a particular community of legal consociates therefore does not necessarily or typically require moral validity according to a universal standard or procedure. It merely means that the "enacted norm has been sufficiently justified and is typically actually obeyed as well" within the relevant community.[32] Given that most legal communities (read territorially bounded nation-states) are complex and culturally diverse, Habermas acknowledges that pragmatic compromises—negotiated agreements that balance interests and accommodate differences—are often necessary since reaching moral or even ethical agreement on the priority of public values and generalizable interests may not always be possible.[33] But the discourse principle—normally oriented toward consensus—is still brought to bear indirectly, *through fair procedures that regulate the bargaining*.[34] In other words, *fair* bargaining and the reaching of fair compromises still presupposes the discourse principle. Accordingly, law making does not entirely displace moral discourses; instead, it incorporates them into its procedures and therefore indirectly into its modes of argument.

Before critically evaluating Habermas's theory it is important to note three important qualifications to this interdependent arrangement among the state, civil society, and the public sphere. First, he argues that the arrangement presupposes a rationalized lifeworld and a *liberal political culture*, something that we would not necessarily expect to find in all state-society complexes.

Second, the "influence" of civil society actors in the public sphere is just that—influence—not political power. In any event, civil society is a "wild" and anarchic complex that resists formal organization.[35] This makes it autonomous and unrestricted but also vulnerable to the exclusionary effects of an unequal distribution of social power (in pointing out this familiar problem Habermas offers no remedies for such a state of affairs).

Third, even in those circumstances where civil society successfully influences the democratically regulated parliamentary assemblies, the instruments of the state (law and administrative power) are seen to "have a limited effectiveness in functionally differentiated societies."[36] Sometimes the state can only steer society indirectly, and must defer to other functionally differentiated subsystems. The political system is to be understood as just one among many social systems rather than *the* apex of all action systems, which means that it that still depends on the successful performance of other action systems (e.g., the economy). More recently, what he calls the "postnational constellation"—the simultaneous pluralization of lifeworlds in the nation and the eroding political autonomy of the state—is reducing the scope for effective democratic politics.[37]

6.3 A Green Critique and Reconstruction of the Habermasian Democratic State

How should critical political ecologists respond to Habermas' highly idealized yet deeply ambiguous reconstruction of democracy, law, and the state? On the one hand, Habermas has tracked the decline of the public sphere yet, on the other hand, he points to its potential to "radicalise communication structures" and "shift the entire system's mode of problem solving."[38] On the one hand, he has argued that the discourse principle is rooted in the structures of linguistic communication in the lifeworld yet, on the other hand, it can only be upheld and realized *within* the framework of a democratic legal system that is necessarily limited in terms of its temporal, spatial, and ethical-existential boundedness and therefore is unlikely to enact truly cosmopolitan legal norms. On the one hand, the democratic state depends for its legitimacy on a vigilant civil society, acting back on the state through the public sphere yet, on the other hand, once the state has enacted democratically determined legal norms, its steering power is limited when considered against the backdrop of other social steering systems. On the one hand, the processes of globalization have contributed to the development of transnational public spheres yet, on the other hand, democracy can only be successfully institutionalized when it is restricted to particular national communities, territories, and administrative structures—something that is difficult to conceive on a regional or world scale. On the one hand, more and more political problems can only be solved at the supranational level, but existing forms of supranational governance will not be able to achieve the levels of political legitimacy enjoyed by territorially bounded, democratic nation-states. We can safely conclude from this overview that the future of democracy looks rather grim.

While it has always been critical theory's concern to pursue its emancipatory aspirations in the context of a sober sociological analysis of the current political order, cynics might suggest that Habermas has raised emancipatory hopes only to dash them or at least mute them to the point of resignation to the familiar liberal democratic domestic order and liberal contractual multilateral international order. Now, in fairness to Habermas, part of his project is to draw attention to the unavoidable and enduring *tensions* between actual political practices and the regulative ideals of democratic law making, or "between facts and norms," as he calls it. His reconstruction of democratic law and politics seeks to show that the law enacted by the modern democratic state "has a legitimating force only so long as it can function as a resource for justice."[39] This is the good news for critical theorists since the moral requirement that power be legitimately exercised is one of the few "weapons of the weak." The bad news, however, is that in practice legitimacy is always a question of degree and the law often falls considerably short of its regulatory ideals not only because of the sway of powerful vested interests but also because of unavoidable and intractable cultural difference and sheer practical exigencies.

Habermas would insist that his idealized claim is not a wishful normative claim but rather a quasi-empirical claim, reconstructed from the implicit presuppositions of communicative action, notably the implicit orientation of actors toward resolving practical disagreements by reaching mutual understanding by means of discursive argument. Habermas's point, then, is merely that all communication is implicitly *oriented* toward reaching mutual understanding by means of reasoned argument rather than coercion, bribery, or bargaining, even if such understanding is not *actually* reached. Moreover such an ideal thus remains a *constitutive element* of every act of communication. Even in highly distorted communicative settings parties can still feel obliged to explain themselves to others by *giving reasons* for their preferred positions if they are to persuade others of the acceptability of their arguments or simply to be recognized as legitimate participants.

Now it might reasonably be asked whether this implicit orientation toward reaching a mutual understanding by means of undistorted communication has always been present in human communication since time immemorial, or whether this is a peculiarly modern expectation—the legacy of the democratic revolution. For our purposes, however, it is not necessary to be detained by the controversy concerning the linguistic foundations of Habermas's theory if it is accepted that democracy, and the principles of political equality and autonomy upon which it is grounded, have indeed become general post-Enlightenment normative expectations or aspirations. In any event, this inquiry proceeds on the basis of this normative aspiration, so the more important question for our purposes is: Given that the deliberative ideal has been shown to be more conducive to environmental justice and reflexive modernisation than liberal pluralistic bargaining, then *how might this ideal be more closely approximated?*

Surprisingly, Habermas does not seriously entertain this question. It is as if he has resigned himself to his sociological analysis of the inherent tensions in and limitations of the democratic state—a state that still bears an uncanny resemblance to the liberal democratic state (notwithstanding his efforts to situate it *between* liberalism and republicanism). Yet accepting this tension ought not to preclude an exploration of how the modern democratic state might be made to function *more* as a resource for justice (including environmental justice) rather than power or mere interest accommodation.

This core question may be probed by breaking it down into two subsidiary questions. First, given that a flourishing public sphere is crucial for the success of the democratic state, then how might its radical potential be furthered? Second, how might the democratic determination of legal norms more closely approximate moral and ethical rather than merely pragmatic modes of reasoning?

The point of both questions is to explore how discursive democracy might be brought into closer alignment with ecological democracy. (We will reserve for the following chapter the question as to whether the procedures of the democratic constitutional state can be made to take into account interests and concerns that transcend the community of legal consociates that make up the nation.)

6.3.1 Realizing the Potential of the Public Sphere

In view of Habermas's analysis of the withering of the public sphere in the face of the domination of the lifeworld by systems of economic and bureaucratic power, and his insistence on the importance of the public sphere to the life of the democratic polity, the question as to how the public sphere might be revived would seem to be *the* most crucial one for critical theorists generally, not just critical ecologists. In *Structural Transformation*, Habermas called for a "march through the institutions," a strategy that recognized the importance of active social forces and an active state in mutually cultivating and maintaining discursive democracy. However, his more recent work is focused on the supposed virtues of abstract democratic law and the networks of discursive communication that law enables and upholds. On this understanding the legitimacy of legal norms is secured by the discursive procedures of democratic will-formation, which in turn are "influenced" by the political opinions formed in the public sphere, understood as the intermediary between civil society and the state. This assumes, first, that the lifeworld is not colonized by systems of money and/or bureaucratic power, and that therefore the public sphere is alive and well and, second, that politicians and state elites are highly responsive to what are presumed to be informed opinions circulating in the public sphere.

While Habermas acknowledges the importance of a flourishing public sphere, he offers no analysis of how this might be achieved other than to underscore the importance of civil and political rights and a liberal, in the sense of pluralist and tolerant, political culture. As John Dryzek explains, constitutional structures are not the only forces that condition, shape, and constrain deliberation in a liberal polity, yet there seems to be no inclination on the part of Habermas to explore how the administrative state or the capitalist economy should be democratized any further. Both are taken as given and "All that matters is that they be steered by law, itself democratically influenced."[40] Indeed, Dryzek pointedly suggests that "Habermas has turned his back on extra-constitutional agents of both democratic influence and democratic distortion."[41]

Now Habermas makes the obvious point that civil society, unlike the state, is not well placed to govern directly, for a variety of reasons, not least of which is that it is a source of discourses that are *critical* of the state. Yet he offers little to reassure actors in civil society that their arguments will be heard, let alone acted upon, by elected representatives and state functionaries. This is most evident in his tracking of the "influence" of political opinion-formation in the public sphere on democratic will formation in the formal legislature, which is merely gestural and surprisingly conventional (i.e., he points to elections and political parties as the major transmission belts of such "influence"). It is as if Habermas accepts that the sway of discourses emanating from the public sphere is both limited and precarious—perhaps more so to the extent that such discourses are critical, free, and unruly—since they cannot all be enacted and therefore must be left to the practical judgment of the leaders of political parties or powerful policy brokers. Yet in liberal democracies operating under the Westminster system where party discipline is strong, the executive typically controls the legislature and exercises considerable discretion and control over the business of parliament, while in the case of the United States, the legislature is often dominated by powerful interest groups. A more critical normative theory of the relationship between civil society, the public sphere, and the state would look to loosening the hold of the political executive and the influence of powerful interest groups on democratic will formation while both deepening and extending the spaces of deliberative policy making and implementation inside and outside the parliament and the bureaucracy. Moreover it is difficult to see how the liberal democratic state might become more of a vehicle for justice than it currently is unless emancipatory social forces both act upon and also march through the institutions of the state and thereby *actively produce* the democratic innovations that are required for less distorted and more critical political communication inside and outside the state. While no one should be so naïve as to expect the tensions between political ideals and practices to be resolved, it is a politically conservative stance not to explore how the gap might be at least lessened.

Habermas's counterfactual communicative ideal is also unhelpful for those seeking to enhance the legitimacy of the democratic state from the standpoint of environmental justice. Of course, Habermas acknowledges that the ideals embedded in democratic procedures are rarely fully realized in practice, but he insists that they nonetheless remain a constitutive element of every act of communication such that the fallible results of political communication nonetheless enjoy "the presumption of rationality." However, since real world political legitimacy is always a question of degree, it is necessary to ask to what extent political communication can always be presumed to be rational in this way, especially when discrepancies between regulative ideals and political practices are recurrent and deep. When it comes to designing constitutions and the procedures of democratic will-formation, a more sober approach might be to draw lessons not from counterfactual ideals or democratic aspirations where human motives are reconstructed as noble but rather from practical political experience, where they are often inscrutable and not always virtuous. Taking a lesson from the neo-Roman republican legacy of constitutional design, the point should not be to presume or assume publicly spirited behavior but rather to design democratic procedures, checks, and balances such that it is difficult for those deliberating, and particularly those exercising public power, to behave otherwise.[42] Moreover, in designing these procedures, checks, and balances, it matters not that office bearers and politicians may be motivated just as much by fear of public shame and a damaged reputation as by the quest for public honor.[43]

Finally, real world political legitimacy can only be satisfied when citizens both believe in, and can practically avail themselves of, their democratic rights by contributing, inter alia, to a flourishing public sphere (or spheres) that have real influence on the state by virtue of the generalizability and persuasive force of their arguments. While Habermas has hinted at the further radicalizing of existing rights, he has not explored what this might entail, nor asked how democratic constitutional states might be made more receptive to good arguments emerging from relatively unconstrained deliberation in civil society. Yet an expansion in the range and scope of environmental due process rights (of the kind discussed in chapter 5) would provide one significant means of deepening and extending democracy in both the state and civil society.

By limiting his attention to subjectless discursive procedures and avoiding any analysis of antagonistic social forces, Habermas has replaced the old Marxist preoccupation concerning revolutionary agency with a reconstruction of idealized modes of argument in order to account for the political legitimacy of the legal norms enacted by modern democratic constitutional states in pluralist, liberal societies. In this respect Habermas offers a *reconstructive* theory of the democratic state, not a *praxeological* theory; that is, he does not explore how the moral and political resources within existing social arrangements might be enlisted for emancipatory purposes. Yet one can only be sceptical of how far any push towards emancipatory change—in both the state and civil society—is likely to succeed on the basis of such a profoundly decentered politics that promises merely that we uphold the traditional repertoire of civil and political rights while respecting the right of others to be different. In seeking to distance his theory from civic republicanism, Habermas avoids any analysis of how the necessary social coalitions might be built that could act as a collective force for emancipatory change in the economy, state and society (including the media). Likewise he avoids any examination of the likely political resistance such coalitions might face from powerful social forces. Instead, Habermas's analysis suggests an emancipatory parodox: giving up the old Marxist and republican ideas of an historic class or collective subject or citizenry is necessary to avoid the smothering of difference by zealous ideologues strategically pursuing the one true political path, yet to embrace fully the idea of the decentered society and public sphere is also to abandon the idea of a counterhegemonic struggle and render extremely difficult the task of building the kind of political solidarity that is needed to transform the systems of money and power that have "colonized the lifeworld" and seen to the withering of the public sphere and the dominance of the state by social forces promoting neoliberal economic deregulation.

I am not suggesting that critical political ecology should revive outdated and problematic Marxist and republican ideas of collective political agency. Indeed, those who have defended explicitly green public spheres have forcefully argued—like most critical theorists—that it should not be governed by a single direction but rather reflect a plurality of values, goals, and interests.[44] Douglas Torgerson, in particular, has gone so far as to suggest that green activists and theorists ought to become less fixated with the idea of a green *movement* (which carries with it a focus on the instrumental achievement of particular goals), and more focused on the idea of a green public sphere as an open series of green discourses made up of fluid, diverse, and changing connections and sites of environmental concern and discourse. According to his Arendtian inspired analysis, green political theorists and green activists should become less instrumental in their approach to politics and more attuned to the value of politics—or open political dialogue/conversation—as *an end in itself*. In effect, while the green movement could possibly survive without debate, the end of debate would spell the end of the green public sphere.[45]

Yet Torgerson also argues that a respect for political pluralism and recognition of the value of the public sphere (green or otherwise) need not and ought not rule out new social movement activity intended to bring about change, especially when the end is to uphold or reinvigorate the public sphere and transform the state by improving the prospects for marginalised groups to contribute to political opinion *and* will formation. As he puts it, "The point, then, is not to dispense altogether with instrumentality in the green movement, but to recognize clearly that instrumentality it is not everything, especially in the best of causes."[46] By the same token, all movements for change are necessarily goal directed, and it would be naïve to reject strategic action directed toward creating more socially and environmentally just social structures, especially in the face of political intransigence on the part of all those who benefit from maintaining unjust political and economic structures. Building coalitions among emancipatory social movements is the most likely way in which any major political mobilization will happen. Introducing strategies of empowerment for all those who have been marginalized in the making of decisions involving ecological risks would seem to be a necessary step toward extending and deepening democratic processes. Just as defenders of minority group rights and a "politics of difference" have exposed the limitations and injustices of the liberal, color-blind constitution, so too critical political ecology can expose the anti-ecological biases in liberal pluralist policy making (as discussed in chapter 4). Moreover, devising and campaigning for special forms of representation and greater access to policy making for those who are systematically underrepresented can help move deliberation closer to the ideal of communicative equality. After all, what Jean Cohen calls a "favorable associative environment" does not always arise naturally.[47] It has to be *actively constructed by the use of public powers*—a move that itself requires deliberation and is therefore invariably contested, particularly by those social forces that stand to lose. Such strategies can be justified in democratic grounds. Indeed, for so long as organized public power is not fully democratic, then environmental justice advocates may sometimes need to resort to strategic rather than communicative action in order to establish the conditions for fair and free deliberation.[48] As Jürgen Haacke has argued, from a Habermasian perspective, "[s]trategic action would appear justified only as long as it seeks to establish conditions that allow communicative rationality to unfold its potential, not insofar as it anticipates the possible outcomes of praxis as discourse."[49] In practice, it might mean enlisting the public power of the state to ensure that powerful social actors—including corporate executives, scientists, politicians, and bureaucrats—are publicly *made to listen* and publicly *made to respond* to community ecological grievances by, say, being called before public hearings. While such requirements would seem to offend the ideal of uncoerced communication, when set in the context of more systemic "background injustices" in terms of unequal access to power, knowledge and resources, they ultimately serve the communicative ideal. The same can be said for political strategies by new social movements that seek to disrupt established political or economic practices that are ecologically damaging in circumstances where these practices do not permit any robust critical feedback from affected parties.

The maintenance of the green public sphere may be more important than the maintenance of the green movement in the long run, as Torgerson suggests, but a diverse array of environmentally concerned actors are also essential to the maintenance of diverse environmental networks. This is hardly an elitist revolutionary vanguard but rather a creative, shifting, and therefore always provisional set of understandings and alliances among new social movements, scientist, research institutes, ordinary citizens, and ecologically modernizing firms whose claims must always be publicly redeemable when challenged.

Within the broad tradition of critical theory, the Gramscian inspired theory of Robert Cox is more sensitive than Habermas's recent work to the role of antagonistic social forces in reproducing and/or challenging hegemonic practices and understandings. Such an approach is also sympathetic with the post-Marxism of Ernesto Laclau and Chantal Mouffe and the ecosocialism of James O'Connor, which see the creative interplay of a diverse range of new social movements as the "new historical subject."[50] This is a less decentered account than that offered by Habermas and Torgerson but considerably more decentered and pluralistic than the traditional Marxist understanding of revolutionary agency. In defending the notion of decentered counterhegemonic strategies by new social movements, Laclau and Mouffe seek to avoid the unpalatable alternative of surrender to the processes of social fragmentation while also acknowledging plurality and the different lines of conflict between different social movements. Of course, as Torgerson has pointed out, even this looser, relatively more decentered concept of counterhegemonic strategies still requires maintaining some notion of friend and enemy, and means and ends in the political struggle for emancipatory change.[51] For Torgerson this is always problematic since it presupposes a privileged standpoint and authoritative answers to normative and strategic questions as to what it should mean to be green and how green goals might be achieved. Hence his preference for open discourse and the spatial metaphors of public spheres and arenas, rather than to heroic historical actors, strategic calculations, and metaphors of movement, struggle, and change. Yet, while no social movement is immune from the dangers identified by both Habermas and Torgerson, social movements would cease to function qua movements if they were to abandon all goal-directed action.[52] The upshot would be the end of any concerted social struggle against established patterns of privilege by means of democratic politics. Emancipatory movements would forever be caught by Habermas's emancipatory paradox in ways that would freeze existing power arrangements. Strategies of empowerment in the form of political mobilization toward the kinds of green discursive designs defended in the previous chapter would enable communicative rationality to unfold its potential and are both necessary and justifiable on democratic grounds. Of course, such strategies must be continuously and discursively justified among the emancipatory social actors if the necessary social solidarity is to be maintained.

Moreover, as Dryzek warns, the public sphere is an empirical category that encompasses the discourses of a whole range of movements and actors from left-liberatarian to right-wing nationalist movements.[53] Thus the public sphere is merely a communication network directed toward the state: it does not, by itself, serve as a normative standard—at least not in its empirical manifestation. Rather, the normative (and critical standard) arises from the implicit presuppositions embedded in, and reconstructed from, ordinary communication in the lifeworld, which is oriented toward mutual understanding. Whereas the bureaucratic state and the market coordinate action in accordance with systemic imperatives, civil society is coordinated in decentered nonsystemmic ways via communicative interaction rooted in associational activities of the lifeworld. However, as Iris Young suggests, it is more helpful to think of state, market, and civil society not as separate realms but rather as different kinds of activities or perhaps rationalities.[54] This also makes it possible to acknowledge, for example, the interpenetration of these activities or rationalities in all three realms. For example, communicative activities take place within the state and the market while bureaucratic and economic rationality can also be found to operate in NGOs.

Nonetheless, Habermas and neo-Habermasians such as Dryzek and Young, all seem to agree that communicative rationality embedded in ordinary, nonspecialized communication in the lifeworld is more sensitive to detecting, identifying, and thematizing new social and political problems (including ecological problems) than systems of money and power, as the history of environmental, anti-nuclear, feminist, and multicultural social movement NGO activity attests. It is precisely because civil society is plural, diverse, voluntary, unpredictable, creative, uncoopted, and decentered that it is able to play this critical oppositional role vis-à-vis systems of money and power.[55] However, as Young goes on to note, it is also because civil society is unruly, uncoordinated, and decentred that it is not able to substitute for the critical functions performed by states or engage in systematic social steering. That so many citizens' and NGO campaigns are directed toward (rather than away from) the state attests to the recognition of the important functions that are uniquely, or most effectively, performed by states.[56] These functions include security, social welfare and income redistribution, economic regulation, the provision of public goods and services (including environmental protection), and the upholding of the rule of law and democracy.

On this view, the state, the economy, and civil society may be seen as both limiting and supporting each other.[57] Although there are tensions between the state and civil society "they are both necessary elements in a democratic process that aims to do justice."[58] This also suggests that social movements and NGOs pursuing social and environmental justice goals should be working on all fronts in mutually supporting ways by, for example, nurturing new identities in civil society, pressing for greater corporate accountability, and pressing for new policies and political practices on the part of the state and its functionaries. Yet the state still stands out as especially crucial in this multipronged process because of its unique capacity "for coordination, regulation, administration on a large scale that a well-functioning democracy cannot do without."[59] It is therefore not just a "necessary evil" but rather an institution that has uniquely positive capacities to curtail the harmful effects of economic power and "potentially and sometimes actually they exhibit uniquely important virtues to support social justice in ways no other social processes do."[60] What is so crucial about the public sphere is that it simultaneously plays the negative role of opposition and critique and the positive role of policy influence.[61] Thus, for Young, "democracy is better thought of as a process that *connects* 'the people' and the powerful, and through which people are able significantly to influence their actions. Democracy is more or less strong or deep to how strong are these connections and how predictable that influence."[62] Habermas's model offers only weak connections (theorized as nebulous influence) and no predictability.

Adapting Young's argument, we might say that the virtuous state should be concerned not simply with laying down a just framework of civil and political rights and discursive procedures that enables democratic participation. It should also be concerned with actively intervening in society and the economy in order to promote social and environmental justice (including wealth redistribution) by means of strategies of political empowerment precisely because they are essential for the democratic self-determination and self-development of the many, not just the few.[63] Such arguments clearly go beyond Habermas's pure proceduralism.

Whereas Young seeks to emphasize what the state can do for democracy, more radical critical theorists such as Dryzek have tended to focus more on how states tend to thwart emancipatory movements and how critical theory might avoid lapsing into an easy accommodation with liberal constitutional thinking by focusing on public spheres as sources of democratic critique and renewal.[64] Now Drzyek, like Habermas and Young, acknowledges that public spheres—however oppositional—only take on meaning and shape in the presence of the state. He also acknowledges that the critical independence and robustness of civil society is crucially dependent on the state, particularly in the ways in which it organizes (or obstructs) interest representation.[65] Thus he concedes that it is impossible to develop a flourishing civil society by turning one's back on the state. However, Dryzek is much more deeply suspicious than other neo-Habermasians, such as Iris Young, of the tendencies of states to seek to coopt and neutralize oppositional movements. Accordingly his primary concern is to clarify those circumstances when it is best for democratic advances to be sought *inside* or *outside* the state. Not surprisingly, he finds no universal answers to this question. From the point of view of new social movements, he argues that these are ultimately strategic questions that are sensitive to time, place, and circumstance. Nonetheless, from the point of view of furthering an oppositional civil society, Dryzek suggests that "benign inclusion" in the state should only be pursued when the movement's goals can be assimilated by the state and when inclusion does not exhaust the discursive or critical capacities of such movements.[66]

Now Dryzek is right to remind us that the conditions for authentic as distinct from merely symbolic political participation in the state are demanding. Moreover political inclusion in the state by civil society actors often runs the risk of depleting the supply and critical independence of such actors in civil society. Indeed, he maintains that the histories of insurgent democratic movements reveals a common pattern of inclusion or absorption and democratic loss according to Drzyek's criteria.[67] He is especially wary of such inclusion when it is sponsored by the state for the state. Dryzek's analysis is based on a deep mistrust of the motivations of governments and state elites, which he sees as favoring state "imperatives" over the goal of democratization.[68] In other words, state elites are understood as having no incentive to promote democratization; to the extent to which they include civil society actors in policy making, such as in corporatist bargains, it is to quell opposition, discipline civil society actors, and promote state imperatives. Even the popular dualist strategy of some movements, which is to promote opposition in the public sphere *and* inclusion in state policy making, is considered by Dryzek to be appropriate and effective only when the movement's interests can be brought into accord with state accumulation and legitimation imperatives—otherwise, one can expect inclusion to be merely symbolic rather than authentic. Ironically, what he calls "passive exclusive states" (i.e., states that do not actively sponsor the inclusion of civil society actors in the processes of policy making) are, by Dryzek's account, good for democracy because they are more likely to prompt a discursively vital and oppositional civil society.[69] For Dryzek, "a truly inclusive state would corrode the discursive vitality of civil society . . . and so undermine the conditions for further democratization."[70]

While Dryzek rightly alerts all critical theorists to the dangers of state cooptation of social movements, his conclusions must be read in the context of his preoccupation with maintaining the vibrancy of civil society and the public sphere against a rather limited conceptualization of the state as "the administrative state" driven by systemic imperatives. Yet state imperatives are not autonomous from civil society and there is nothing fixed or inevitable about their functions and goals. Rather they are produced and reproduced by the relationships and understandings that are forged, inter alia, between state and civil society actors. So-called state imperatives are, after all, merely reified social relations, practices, and understandings. The point, then (contra Dryzek), is to challenge those relationships, practices, and understandings that are not environmentally inclusive and to work toward making the settings in which they are forged less distorted from the point of view of ecological democracy. To the extent to which this occurs, the zone of policy discretion can be expected to open up (or at least be less "imprisoned"), and we can also expect the goals and functions of states to change. Dryzek himself notes that state imperatives have changed, or rather expanded, over time in response to societal problems and the claims of social movements, and nowadays it is possible to recognize environmental conservation as a new state imperative, at least in the form of ecological modernization.[71] However, these must be ultimately understood as *political and socially negotiated* changes involving political contestation, not autonomous changes in response to objective systemic imperatives.

Whereas Dryzek's overriding concern is to maintain a vibrant civil society and public sphere, the concern in this inquiry is broader: How might we enhance state reflexivity or ecological problem-solving capacity? This is the same as asking: How might the communicative context become less distorted and more inclusive? Dryzek's empirical observation that "passively exclusively states" tend to prompt oppositional movements should not be taken as an argument for avoiding engagement with the state, although it does warn of the dangers of unthinking strategies of inclusion. The point, as I see it, is to make the democratic state more responsive to such critical feedback, acknowledging the crucial role played by civil society actors and public spheres in the processes of problem detection. The goal should not be to eliminate the unavoidable and necessary tensions between civil society and the state. Rather, it should be to explore how they might be played out in more creative ways, particularly for those groups that have historically been excluded or marginalized in the processes of policy making.

6.3.2 From Pragmatic to Moral Deliberation (and Back Again)

Although Habermas sets out to distinguish discursive democracy from republican and liberal accounts, he ends up defending an arrangement that has many of the hallmarks of conventional liberal pluralism, where pragmatic reasoning predominates over moral and ethical reasoning in the policy making and legislative processes. His reassurance that pragmatic reasoning is nonetheless grounded in rules of fair bargaining that ensure equal consideration of the interests of all parties is unlikely to satisfy those who are concerned about unequal communicative power in the real world and "unrepresentative representation" in government in terms of discourses and social actors. Nor is it likely to reassure social and environmental justice advocates who find it desirable that political opinion and will-formation aspire toward higher rather than lower levels of intersubjectivity. Ecological democracy aspires toward moral rather than pragmatic reasoning, since moral reasoning directs deliberations toward the widest possible constituency of affected parties. Our ambit claim for ecological democracy is one that, ideally, asks participants in any discursive forum to examine proposed norms from the perspective *all* significantly affected others, citizens or noncitizens, the living and the not-yet-born, human and nonhuman. The quest is to develop procedures and decision rules that might encourage participants to at least strive to adopt "the moral point of view" rather than *merely* seek to further their own narrowly conceived interests by strategic bargaining. (I say "merely" since the particular interests of participants should not be excluded from discursive dialogue.)

Given that the legitimacy of Habermas's entire constitutional and political order turns on the supposed fairness of the conditions and procedures of political communication, these questions surely demand closer attention. For his part, however, Habermas merely offers an idealized reconstruction of moral, ethical, and pragmatic *modes of argument* (focusing on the levels of intersubjectivity required); he does not actually explore how practical reason might achieve higher levels of intersubjectivity so that decisions made in the name of the state advance justice rather than material interests. Nor is the prevalence of pragmatic interest accommodation likely to build the kind of social solidarity that is necessary for common problems to be dealt with in common ways. Merely arguing that solidarity ultimately resides in the linguistic bond that holds together each communication community avoids the difficult issue of different languages and different communicative genres and competencies in multicultural societies. Here I focus on two interrelated levels on which these failures are problematic from the perspective of critical political ecology. The first is on the scope of the rules of discourse and what counts as "the moral point of view," and the second is on pragmatic compromises, when they may be necessary and when they should be avoided or overridden. In short, Habermas's account of moral reasoning is not moral enough while his acceptance of pragmatic compromise is too complacent and uncritical.

Habermas's proceduralist defence of discursive democracy, like liberal accounts of justice, draws a distinction between questions of morality and the good life. Questions of morality (or justice) concern who are morally competent subjects; they set the ground rules of the dialogue and shape the conversation in ways that prompt participants to respect and take into account the situation of all linguistically competent agents.

Questions of humanity's relationship to nonhuman nature (with the exception of certain nonhuman mammals) are considered *ethical* questions to be determined in real discursive dialogue; they are not matters that ought to condition or constrain the rules of discourse in advance of the dialogue. However, just as liberal accounts fail to acknowledge the unavoidable link between questions of the right and the good, Habermas also fails to acknowledge the ways in which the moral norms underpinning different discursive protocols can shape and constrain the substance of ethical dialogue.

For critical political ecology, the question of what sort of beings beyond the human realm can be represented and considered in dialogue is a matter of justice (who or what matters morally, and who or what is entitled to be represented) not mere ethics on Habermas's understanding of these terms. Treating these questions merely as ethical questions necessarily leaves the fate of future generations and nonhuman others to the uncertain motivations and inclinations of those who happen to be present in any given dialogue. Accordingly no systematic consideration of nonhuman others can ever be expected, least of all guaranteed, by the discourse ethic. However, if environmental justice considerations were to be incorporated into discursive protocols, then participants would be required at least to consider the plight of environmental victims. Again, there can be no guarantees that all potential environmental victims will be protected—these are questions to be determined in real dialogue—but the likelihood of risk minimizing and risk averse decisions is nonetheless considerably greater. Habermas's argument that the democratic constitutional state presupposes a *liberal political culture* makes it clear that his account of democracy does not seek to make any major inroads into conventional accounts of liberal autonomy, which assumes that only humans matter, and therefore presupposes an instrumental human posture toward the nonhuman world.

Habermas's uncritical acceptance of a liberal political culture is also manifested in his complacent acknowledgment that pragmatic arguments and agreements are likely to predominate over moral and ethical arguments and agreements in the negotiation of legislation. However, I want to suggest criteria for determining when this might be acceptable and when it might be suspect from the standpoint of ecological democracy.

In cases of intractable moral disagreement stemming from deep-seated differences in philosophical, religious, or cultural framewords, pragmatic compromises appear to be the only way in which environmental policy deadlocks can be resolved. Indeed, the new school of environmental pragmatist thought has argued that it is important that environmental democracy be open-ended, that participants be respectful, open-minded, good listeners who are prepared to work creatively with the moral and cultural resources at hand. Sometimes this may mean merely seeking only the minimum necessary common ground for the purposes of instrumental environmental problem solving. In this context the tactful avoidance of deep-seated moral, religious, cultural, and social differences and the searching out of pragmatic solutions to practical problems is more productive than allowing unnecessary and endless heated debate about deep-seated environmental values and cultural and philosophical differences. For environmental pragmatists it is important that the procedures of democratic deliberation be radically open in order to leave the clarification of issues, agenda setting, practical problem solving, and adaptive learning and management to real stakeholders who constitute the relevant "community of inquirers" that must live with, and learn from, the consequences of their decisions. Under these circumstances compromise, incremental change, and even "muddling through" are preferable to holistic social engineering, which is likely to be particularly insensitive to cultural difference.[72] In this respect environmental pragmatists follow the Popperian tradition according to which "holistic engineers" are the "enemies" of the open society.

However, when we turn to situations that predominantly involve the clash of material interests and "power trading," different considerations ought to come into play. Under such circumstances the radical indeterminacy of democratic procedures and agendas can provide a context for the sway of powerful interests over less powerful ones. Moreover the instrumental, problem-solving approach of environmental pragmatism runs the risk of being too accommodating and therefore not critical enough of the existing constellation of social forces that drive environmental degradation. In short, the more radically open are democratic procedures, the more they are susceptible to abuse as well as good use in situations where there are significant disparities in social, economic, and communicative power. Under these circumstances, insisting on environmental due processes of the kind outlined in the previous chapter would ensure at least that systematic consideration be given to excluded voices without insisting, a priori, on particular outcomes. Of course, "controlling the sway of power"—whether it be economic, social, or cultural power—is no easy task, and certainly not something that can be managed *merely* by good constitutional design. Nonetheless, the kinds of constitutional provisions discussed in the previous chapter—notably the entrenchment of the precautionary principle—would certainly go some way toward instantiating ecological democracy and therefore making the discursive context at least more conducive to the enactment of more environmentally sensitive public policies.

There are some parallels here in the political debate about the need for a politics of recognition. Habermas's response to the multicultural question is to defend "constitutional patriotism" over ethnic nationalism. This also entails a politics of recognition and a considerable effort on the part of dominant ethnic groups to refrain from assuming that their culture and identity represents the national culture. By the same token, critical ecologists are seeking a politics of recognition of nonhuman others along with an effort on the part of human agents to refrain from assuming their mode of autonomy in the world represents the only mode of unfolding or being in the world. Extending the politics of recognition to nonhuman agents that do not necessarily conform to recognized modes of human agency is both desirable and necessary if critical political ecology is to transcend the purely instrumental posture toward the nonhuman world that is characteristic of liberalism and the more technocentric expressions of ecological modernization.

Even with my recommended procedural safeguards in place, one would not necessarily expect consensus. However, insofar as real world deliberation falls short of the deliberative ideal, critical political ecologists would want this to be from genuine moral disagreement or a mutual acceptance of practical exigencies rather than from the sway of economic, social, or cultural power. Thus, while pragmatic accommodation may sometimes be necessary or unavoidable to reduce conflict, reach decisions, and stabilize social relations, it is necessary that the variable of "distorting power" be controlled as much as possible if green democracy is to aspire toward genuine inclusiveness. Although Habermas distinguishes between different types of practical reason, he makes no distinction among these *different types of pragmatic compromise*, nor offers any criteria for evaluating what might be acceptable compromises.

This chapter has sought to build on the work of Habermas and neo-Habermasians such as Young and Dryzek by exploring how the democratic constitutional state might be transformed in ways that maintain an optimal relationship (understood as a productive tension) among civil society, the public sphere, and the state. In general, this requires the further democratization of the state, civil society, and the economy so that the system imperatives of the state and the economy are more firmly constrained by the needs of the lifeworld. The green democratic state should seek to uphold the ideal of an inclusive and outward-looking democracy that seeks to facilitate the reaching of provisional (and therefore revisable) common understandings that are tolerant of difference. However, it will invariably be the case that not all interests or conceptions of the common good, and not all particularistic standpoints can be equally accommodated. In cases of intractable conflict, the discursive norms and procedures upheld by the green democratic state ought to favor the interests of the dominated over the interests of the dominators, provided that any intervention leads to less rather than more domination.

The project of building the green state of the kind I have defended can never be finalized. It must be understood as an ongoing process of finding ways of extending recognition, representation, and participation to promote environmental protection and environmental justice. Moreover such a project must also entail exploring to what extent the territorially and legally delimited green democratic state might be able to serve as a vehicle for environmental protection and justice at home *and* abroad. To this end, the following chapter seeks to negotiate some of the enduring tensions between the bounded nature of state democratic governance and the unbounded nature of green morality.

7

Cosmopolitan Democracy versus the Transnational State

7.1 Principles of Democratic Governance: Belongingness versus Affectedness

The previous chapter sought to clarify the mutually dependent relationship between the green democratic state and green public spheres. However, public spheres, especially green ones, are fluid, wide-ranging, and not confined to the discursive spaces of parliament, the state or the even the civic nation but rather stretch to encompass discourses of local, regional, international, and global common ecological and social concerns.[1] Habermas's *ideal* communication community is likewise not in any way bounded by national communities or territories since it must reach out to include all those potentially affected by proposed arguments and norms. The ambit claim for ecological democracy extends this ideal by requiring that all those potentially affected by ecological risks (human or nonhuman, present or future generations) should have some meaningful opportunity to participate or otherwise be represented in the making of the policies or decisions that generate such risks. From the perspective of critical political ecology, the achievement of public-spirited environmental regulation presupposes the incorporation of such transboundary discourses of public interest. The question explored in this chapter is how far the state might serve as vehicle for transboundary democracy by facilitating ecological citizenship both within and beyond territorial borders. In short, to what extent might the green democratic state emerge as a legitimate "transnational state," that is, one that enjoys the confidence of its own citizens *as well as* other communities that it may serve or assume responsibilities toward? Such a question blurs the distinction between, and challenges the assumptions of, mainstream international relations theory (of the kind examined in chapter 2) and domestic political theory. As I show, the question is neither futuristic nor fanciful.

The previous chapter showed that Habermas's sociological analysis of *formal* democracy is one that is necessarily anchored in the nation-state, or else a supranational constitutional structure that mimics the nation-state, defined by territorial borders and a national or supra-national community bounded by sentiments of "constitutional patriotism." That is, as soon as the discourse ethic is institutionally anchored in the nation-state and legal system, the unlimited communication community shrinks back to the particular community of citizens as legal consociates of particular states, and the formal deliberators and decision makers shrink back to the political executive and the democratically elected representatives sitting in the legislature (who may or may not be swayed by arguments in the public sphere). So, while unrestricted democratic deliberations within the open-ended public sphere may take on a transnational or global rather than purely local or national form, Habermas's defense of the constitutional state offers no formal, institutionalized means to recognize, or otherwise incorporate, the concerns of those living outside the nation-state, even though they may be materially affected by decisions made within the nation-state. This is because institutionalizing the discourse ethic for the purposes of the enactment of *legal norms* necessarily restricts the deliberative community to the community of citizens within nation-states. Citizenship is restricted to members of the nation-state, understood as a community of "legal consociates." The legitimacy of democratically determined legal norms requires that the authors of the law also be the addressees of the law. So while the moral and critical core of Habermas's theory of discursive democracy remains thoroughly cosmopolitan insofar as it asks participants in any discursive dialogue to judge proposed norms in terms of how they affect others—regardless of membership in any particular bounded community—his theory of law, legitimacy, and the state ultimately takes on an unavoidably communitarian hue insofar he believes that the practical reason that produces legal norms cannot transcend the particular culture of particular communities.

This apparent tension in Habermas's theory of discursive democracy captures a broader tension in democratic theory and practice between two quite familiar but very different principles that seek to ground the rights and entitlements of individuals (and in some cases groups) to participate in or otherwise be represented in democratic decision making. The first of these is the cosmopolitan principle of "affectedness," which may be traced to Kant, while the second is the communitarian principle of "belongingness" or "membership," which may be traced to ancient Greece. The cosmopolitan principle of affectedness is a universal principle, grounded in human reason (or, in the case of Habermas's ideal communication community, the implicit norms of human communication which are "reconstructed" by human reason). It is ultimately concerned to uphold the autonomy of each and every individual by insisting that persons should not be bound by norms that potentially affect them if they have not given them their free and informed consent. Such a principle is understood to be applicable to all individuals, regardless of their membership in particular communities.

The communitarian principle of belongingness or membership, in contrast, restricts participation to those who belong to, or are members of, a particular community or demos, and is essentially understood as a collective project that is concerned with the self-realization of individuals *qua members of a particular community*. Indeed, the word *deme* in preclassical Greece meant the territory inhabited by a tribe.[2] Thus democratic rights and entitlements are restricted to those who are members of the "tribe," regardless of who else might be affected by decisions made in the name of the tribe. For Michael Walzer, the overriding communitarian principle of democratic governance ought to be "self-determination of the tribes."[3] Individual autonomy in this context is realized not by the possession of abstract rights but rather by participation in the collective life of the tribe. (There are different nuances associated with the concepts of belongingness and membership insofar as the former is primarily concerned with questions of identity, and tends to have a "thrown quality," while the latter can encompass membership and participation in an association, which suggests an element of choice. I will return to these differences below; for the moment the concern is merely to draw out the tensions between the two general principles.)

The principles of affectedness and belongingness/membership each provide a conceptually distinctive basis upon which to institutionalize democracy. Ideally, at least, the cosmopolitan principle of affectedness applies to all citizens of the world who are affected, regardless of membership in any particular community, while the communitarian principle of belongingness or membership requires the delimitation of particular political communities with clear territorial boundaries for the purposes of democratic governance. Of course, in practice, administrative boundaries of some kind are just as necessary for cosmopolitan governance as for communitarian governance to enable the marshaling of democratic will and the implementation of democratic decisions by means of law. In both cases, then, determining the boundaries of the relevant political community is always a contested matter and in practice one usually finds a blending of the principles.[4] Relatively speaking, however, there appear to be many more practical difficulties associated with delineating the relevant community of affected individuals for the purpose of institutionalizing cosmopolitan democracy than there are with delineating the relevant community and territory for the purpose of institutionalizing communitarian democracy. Yet even delineating the boundaries of national or ethnic groups and associated territories for the practical purposes of implementing the communitarian principle of self-determination is rarely a straightforward matter. Although Habermas has sought to situate his theory of democracy *between* classical liberal (cosmopolitan) and republican (communitarian) accounts, his "communicative communitarianism" appears as a "second-best" solution to his cosmopolitan communicative ideal as a mode of democratic will formation.[5]

Given that the ambit claim for ecological democracy is clearly a cosmopolitan ideal, it is necessary to explore to what extent it might be practically realized and how it might stand in relationship to the communitarian principle of collective self-determination. To this end, I will critically explore Habermas's argument that democratic will-formation must necessarily be delimited in the way he suggests. From this critical encounter I will suggest how discursive democracy might be repositioned in more creative ways between conventional civic republican accounts that resolutely defend the virtues of the national community as the proper locus of democracy, and (in this case) liberal accounts of global cosmopolitan democracy (exemplified in the work of David Held), which regard national boundaries, belongingness, and national membership as increasingly arbitrary or irrelevant to democratic decision making in a globalizing world.[6] The argument I defend in favor of the "transnational (green) state" will be found to be more ambitious than Habermas's account of democratic will-formation in the nation-state but much less ambitious than Held's global cosmopolitan democracy. Such an argument will enlist and enlarge Habermas's defence of "constitutional patriotism" to show how the principle of membership *might be extended,* rather than obliterated, by the principle of affectedness in ways that enable states, understood in this context as legal steering systems, to serve not only the national community but also other kinds and layers of communities in circumstances where a significant ecological nexus can be found.

As a preface to this discussion, it must be reiterated that Habermas is not exclusively fixated with the nation-state. Indeed, he clearly welcomes the development of postnationalist democracy.[7] For example, he looks forward to the day when social and environmental policy have expanded to the same point as economic and monetary policy in the European Union, leaving nation-states to deal only with those matters that do not generate spillover effects.[8] Only then does he consider that democracy will be able to "catch up with markets" in Europe. And catch-up it must if it is to produce the kind of common policies that are able to tackle problems that cannot be adequately managed at the national level. Yet Habermas does not consider that the emergent processes of postnational democratic-*opinion formation* in the European Union have been able to connect with the processes of European democratic-*will formation* in ways that provide anything like the legitimacy that European social democratic nation-states seemed to enjoy in their heyday. Consistent with his analysis of democratic law, Habermas argues that the democratic deficit in the European Union can only be bridged when citizens share a sense of European identity and see themselves as the authors and addressees of a truly *European* law. This requires the transition from intergovernmental agreements to a common political existence, entailing a common constitution and a shared commitment to democratic procedures that can enable common political opinion- and will-formation at the European level.[9] This also presupposes a pan-European media, public sphere, party system, and a common language (this will have to mean English, which is a second language for many European citizens). In effect it seems that only when the EU itself has replicated all the essential trappings of the nation-state that it will have the requisite legitimacy to enact supranational laws within the regional territory of the EU. In this respect the post- or supranational constellation defended by Habermas in the EU might be described as a "meta-national constellation." That is, the features that contribute to the legitimacy of the nation-state have been hitched up to a more abstract level to form something like the "United States of Europe," bound together by a common identity of the sort that Habermas believes can ground the necessary redistributive politics to tackle common problems such as ecological degradation and unemployment.

To date, however, a shared commitment to Europe has not been able to emerge with sufficient force to counteract the diminution of popular sovereignty in the nation-state or to replace this diminution with new forms of governance that can match the levels of legitimacy hitherto enjoyed by democratic nation-states.[10] Habermas considers that the prospects for the development of a truly global, as distinct from merely European, postnational democracy are even more remote. Nonetheless, he singles out as the most promising development at the global level the emergence of transnational networks of communication by nongovernment organizations (NGOs) and international organizations (IOs) operating in and around multilateral treaty making. However, Habermas still considers that these networks do not even come close to matching the legitimacy requirements of democratic nation-states.

Despite Habermas's modernist, cosmopolitan aspirations, then, his analysis proceeds on the footing that there is an (almost) unbridgeable gulf between domestic and world politics, such that a "regional domestic politics" in the EU is probably the best we can hope for. Habermas's claim is that for any society to act back upon itself *as a collectivity*, it must be delimited since the idea of democratic "self-control" presupposes a clearly delimited "self" or society that can enact positive law in relation to a defined territory. For Habermas, "the modern territorial state thus depends on the development of a national consciousness to provide it with the cultural substrate for a civil solidarity."[11] Local, personal bonds are extended to more abstract bonds among citizens of the nation. Although remaining strangers, citizens nonetheless feel a sense of shared identity, social solidarity, and responsibility with other citizens, at least enough to made significant sacrifices, such as by paying taxes for public purposes or going to war in defense of the nation.[12] Moreover, what he calls the "dialectic of legal equality and factual inequality gives rise to the social welfare state, whose principle goal is to secure the social, technological, and ecological conditions that make an equal opportunity for the use of equally distributed basic rights possible."[13] Thus Habermas insists that democratic self-rule demands enclosure, whether the boundaries of the "self" are national or European. The idea of citizens, as equal consociates under law, must go with a territory and an administrative state (necessary to carry out the democratic will by means of positive law), to make up the entity known as "the nation-state" (or the supranational-state). Effective social steering, which is to say the enactment of legitimate law, demands all of these things. The democratic problem, for Habermas, is that national solidarity is eroding at the same time as the state's capacity for social steering has been diminished by the processes of globalization.[14]

Habermas's response to the erosion of social solidarity stemming from globalization and the increasing movement of peoples is to argue that the connection between citizens in modern multicultural polities should no longer be based on ethnicity but rather "constitutional patriotism" or a shared commitment to democratic procedures that enable *abstract, legally mediated* social integration. Indeed, Habermas now badges his approach as a "communicative account of republicanism" which he argues is superior to ethnonational or traditional communitarian conceptions of the nation.[15] In terms of the division in communitarian thought foreshadowed above, such a conceptualization of citizenship opens itself to "the inclusion of the other," since it is based on the virtues of *membership* in a legal association rather than the virtues of *belongingness* to a particular cultural or ethnic identity or "tribe." As Habermas explains, this procedural and legalistic understanding of legitimate rule "connects sovereignty with the private and public autonomy granted everybody equally *within* an association of free and equal legal subjects."[16] The great virtue of Habermas's constitutional state is that it can open itself internally to a range of different cultural and social identities by welcoming all citizens to participate equally in the civic nation. The downside, from the point of view of critical political ecology, is that it formally shuts itself off from the outside. In this sense it is internally cosmopolitan but externally communitarian.

Yet Habermas's claim that political will-formation must be bounded or restricted to the national community does not stand up to critical scrutiny. Habermas seems to have unnecessarily overcoded the boundaries of the nation-state and its associated legal system by judging contemporary developments against a highly idealized and somewhat nostalgic schema of the post-1945 western European nation-state. He also seems to have underestimated the potential for a vibrant public sphere and *innovative discursive procedures* to lift the horizons of not only democratic opinion formation but also democratic *will-formation* beyond the territorially bounded national community of citizens. After all, if the ties that bind citizens in a discursive democracy are ultimately based on a shared commitment to *democratic procedures* rather than ethnicity or nationality per se, then there is no moral or practical reason why innovative transnational democratic procedures cannot extend these ties beyond the national community. I will seek to show that to the extent that democratic rights and responsibilities can be legitimately and effectively transnationalized by multilateral cooperative arrangements among states, then restricting democratic membership to "legal consociates" of only one national community is arbitrary.

In liberal democratic states, the "democratic will" that arises and is converted into law is ultimately only an *artifact* of the assumptions, regulative ideals, policy processes, law, procedures, and decision rules that seek to underpin, justify, measure, discover, and implement it. It is not something objective or "natural" that awaits discovery. Similarly the democratic will in green democratic states would likewise be an *artifact* of green democratic procedures and decision rules, informed by green ideals. Within this green normative framework (which is an extended Habermasian framework), the practical task is to discover how the community of legal consociates can be stretched in space and time to accommodate the variable, uncertain, and complex character of ecological problems.

7.2 Communitarian or Cosmopolitan Democracy?

Now civic republicans and other communitarians would doubtless object to the idea that the mere fact of being *potentially* affected by a proposed decision is sufficient to confer citizenship (or even a more limited right to democratic participation within the polity) on foreigners. For civic republicans, national communities provide the most appropriate "self" to ground and activate the principle of self-determination. For David Miller, democratic self-determination is a *collective* goal of a community that presupposes a degree of mutual trust and reciprocal recognition based on a common language and cultural identity.[17] These social bonds make it possible to pursue common goals in ways that transcend individualistic and sectional interests. Thus questions of identity and belonging are understood to be ontologically prior to, and therefore determinative of, questions of democratic design since the principle of self-determination is thought to presuppose a *preexisting self*, understood in collective rather than individualistic terms. National communities are thus preconditions for, rather than historically contingent expressions of, democratic rule.[18] While Miller emphasizes the active and evolving character of national communities (e.g., outsiders can join this community of descent and citizenship can therefore be acquired by adoption)[19] it is the preexisting thrown ties that provide the basis for the active fashioning of the political community.

Yet not all communitarians have taken the national community as the most basic community for enacting self-determination. For Walzer, the right of self-determination should belong to cultural and linguistic groupings that he calls "tribes" (which may exist as subnational communities) while the right of self-determination may be fulfilled by means other than the formation of new states.[20] Similarly Will Kymlicka has argued that democratic politics works best "in the vernacular," that is, in those political communities that share a common language and mass media.[21] This may or may not correspond with the boundaries of the nation-state. Nonetheless, I will accept the communitarian argument that social solidarity of some form is necessary to ground democratic politics. The question remaining, however, is what kind of social solidarity, how can it be cultivated, and how might it be extended?

Now communitiarians might argue that the types of social bonds that characterize transnational communities (e.g., communities of environmental scientists and nongovernment environmental organizations) that coalesce around particular ecological problems are in some significant sense different from and more tenuous than those that characterize national and/or linguistic communities. Such transnational communities are merely voluntary and selective and therefore lack the "thrown character" of national communities. While individuals and groups might *choose* to become members of the political communities represented by organizations such as the World Wide Fund for Nature or Greenpeace International, citizens are usually born into (and usually die in) national or linguistic communities that carry a shared history, language, and public culture that provide an enduring, collective identity, even if these are invariably contested by different social forces within the community. Thus Kymlicka bluntly declares that transnational *activism* is not the same as democratic *citizenship*, which presupposes a common language.[22] (But, if citizenship is understood to be an activity, rather than something that one possesses, then it is possible to dispute this claim, as I show below.) Perhaps the strongest argument to be drawn out of this communitarian understanding of democracy is that it is the unavoidable and continuing character of linguistic social bonds that enables the development of *societal learning*. In contrast, it might be said that political communities that merely coalesce around particular, transnational or international debates or problems are occasional and transient political communities where the prospects for collective social learning and hence mutual understanding can never be as deep or lasting. Yet social learning and mutual understanding can also develop within transnational communities, particular those that have been working on collective problems for a long period of time (e.g., global warming). I will return to this point shortly.

For thoroughgoing cosmopolitans such as David Held and Daniel Archibugi, however, national, linguistic, or cultural boundaries should have neither moral nor legal significance. Thus there are no grounds for privileging or restricting democratic participation to those who happen to live within the boundaries of *any* community since the core question in any democracy should revolve around who is affected by decisions, regardless of cultural or social background or geographic location. Unlike Habermas, Held's radical liberal idealism presses the cosmopolitan argument to what he sees as its logical institutional conclusion by seeking to realign and subordinate states within an overarching global cosmopolitan law, or "democratic public law."[23] Such a realignment is defended as enabling all individuals (as citizens of the global polity) to share a "common structure of political action" understood as "a cluster of rights and obligations which cut across all key domains of power, where power shapes and affects people's life-chances with determinate effect on and implications for their political agency."[24] Only then can power be held accountable wherever it is located—whether in the state, the economy or cultural sphere. These reforms follow Held's analysis of the ways in which the processes of globalization have enabled sites of political, economic and cultural power—including states—to become increasingly *disconnected* from the consequences of exercises of such power. Globalization has facilitated *the restructuring of power relations "at a distance,"* breaking down the assumptions of congruence between peoples, territory, and states.[25] Global democratic public law is intended to enable citizens to confront such power with its consequences and to obtain appropriate political and legal redress. Moreover all power centers and authority systems—including states—would be legitimate only to the extent that they upheld this overarching democratic law.[26] Individuals would be able to enjoy multiple forms of citizenship at the local, national, regional, and global levels. Such a global law would not displace states but rather make them subordinate to an overarching democratic law. As Held explains, "The cosmopolitan model of democracy is the legal basis of a global and divided authority system—a system of diverse and overlapping power centers, shaped and delimited by democratic law."[27] Each of these layers of political community would have limited jurisdiction according to a set of filter tests ("extensiveness, intensity, and comparative efficiency") that seek to implement the principle of subsidiarity, which Held interprets primarily in accordance with the "affectedness" principle.[28] In cases of dispute, "issue-boundary forums or courts" would be established to determine which level of decision making should have jurisdiction, again in terms of who is most affected.[29] It is noteworthy that the majority of examples that he gives to illustrate the benefits of cosmopolitan democracy are environmental examples. He also outlines a program of reform of the United Nations Assembly (including the creation of a Citizens' Chamber), the creation of a new global parliament, an interconnected global legal system, complete with a Court, an international military, and a guaranteed basic income.[30] In effect Held's project seeks to build what Habermas believes is impossible: a "global domestic law" to catch up with the social and ecological problems generated by globalization.

Habermas's accounts of democratic will formation is clearly much closer to Miller's civic republicanism than Held's cosmopolitan democracy. Common to Miller and Habermas is an emphasis on social bonds of the kind that can support a redistributive politics that can secure collective well-being and common responses to common problems. However, whereas Miller's account rests on a set of ontological assumptions about the benefits of membership in a *national* community as a common *cultural community*, Habermas's account plays down the cultural dimensions of national community and seeks instead the cultivation of social solidarity on the basis of a more abstract commitment to democratic procedures that can unite disparate ethnic groups. He sees this as desirable not only to avoid the perils of chauvinistic nationalism but also to acknowledge how the increasing movement of peoples has changed the character of many national communities in ways that challenge the notion of a national community having a singular character (as if it ever did). In short, the social bonds within nation-states have "thinned out" in response to migration and the social effects of economic globalization and therefore need to be replaced with a more inclusive "constitutional patriotism" based on shared *membership* of a democratic legal order rather than cultural community.

Miller also assumes that trust and mutual recognition are preexisting features of national communities that ground the argument for restricting citizenship to such communities. However, here I would side with Habermas's account, which emphasizes the important role played by the democratic legal system in helping to *cultivate* social solidarity that might otherwise be lacking in culturally heterogeneous societies. Habermas's account thus emerges as more forward looking than Miller's and more sensitive to cultural difference and the processes of cultural transformation within contemporary nation-states. Although Miller stresses the active, changing character of national communities, he argues that we must work with the trust and solidarity that can be found in preexisting communities while also assuming that national communities are the only or best communities that can provide such trust and solidarity.

Yet, if the communitarian point is to search out those communities that share trust and solidarity of the kind that gives rise to a preparedness on the part of members to make sacrifices for the common good, then national communities are clearly only one such community. Against Miller, Walzer and Kymlicka have argued that the borders of nation-states rarely coincide with, or are as strong as, the kinds of linguistic, ethnic, and religious bonds that make up many communities residing within or straddling across nations-states. This partly explains why Habermas favours a more abstract social bond to bind the nation that can be shared by different cultures and communities, namely a deep-seated commitment to democratic processes that can enable the development of mutual understanding and civic solidarity. Yet, if social solidarity is to be built upon such a shared commitment to addressing collective problems by means of democratic procedures, then it also seems arbitrary for Habermas (like Miller) to restrict such democratic procedures to *national* communities as if national communities provide the only basis for cultivating democratic solidarity. If national communities can be inclusive internally, then why must they always be exclusive externally, especially in an age of globalization? Indeed, the increasing pluralization of lifeworlds brought about by, among other things, new communications technology and the movement of peoples is now posing a serious challenge to the notion that "the nation" should serve as the exclusive political and cultural community for the purposes of harnessing state democratic power. Ironically the social bonds that unite transnational new social movements that have come together over common, transboundary ecological problems and risks *have been created out of discursive processes* (albeit informal ones) and are therefore likely to possess exactly the kind of "patriotism" toward democratic procedures that Habermas defends. Here ecological citizenship is not so much a status that one possesses, it is a shared activity, united around collective problems.

Moreover, for every argument typically advanced in favor of national communities as the exclusive locus of democratic citizenship, one can also find countervailing arguments or at least a new set of challenges that weaken the case for understanding democracy exclusively in terms of national self-determination. For example, going to war in defense of the nation and paying taxes for the common purposes of the nation are typically taken as evidence of the special kinds of bonds that prevail in (cohesive) national communities. Now it is certainly true that major security threats can have a galvanizing effect on the building of national solidarity, prompting major sacrifices in defense of the nation.[31] However, joining the military when conscripted and paying taxes are compulsory and do not necessarily indicate a willingness on the part of citizens to make voluntary personal sacrifices for the nation. Moreover it is during major security threats that national bonds can become particularly vulnerable to manipulation by political elites for "reasons of state" rather than reasons of democracy. At the same time a range of new ecological security threats have undermined the concept of territorial defence while economic globalization has created new divisions between "the haves" and the "have-nots" that transcend national communities.

It is also true that states need revenue to function and that extracting taxation from the national community presupposes that the revenue will be returned to the national community and not to strangers who are not liable to pay taxes. Yet citizens who do not earn enough to pay taxes are not thereby denied their entitlement to state benefits (although the Third Way rhetoric of mutual responsibilities perhaps points in this direction). That is, the *moral* criterion for redistribution *within* the national community is ultimately need rather than the capacity of individuals to make a financial contribution to the nation. Moreover, as moral cosmopolitans such as Charles Beitz and Thomas Pogge have argued, the breaking down of economic and social borders strengthens the case for redistribution between the haves and the have-nots between states, not just within states.[32]

None of the foregoing arguments are meant to deny the significance of national communities. National bonds may be imaginary, but this makes them no less palpable, particularly in times of World Cup soccer games, war, and national disasters (e.g., floods and fires) where personal sacrifices by fellow nationals for fellow nationals in dire straits can take on heroic proportions. Moreover Habermas is also right to argue that social steering systems cannot be divorced from the particular, existential communities they are meant to serve. However, my central point is that there is no good reason why such steering systems must be *exclusive* to only one kind of existential community at all times. National communities are only one kind of community and they are under increasing strain from the processes of globalization. If nations are imaginary communities based on abstract rather than embodied social bonds, then there seems to be no good reason for denying the significance of other kinds of imaginary communities that come into being in response to common problems that transcend national boundaries or simply in response to human suffering or ecological degradation wherever it may occur in the world. The making of *voluntary* donations to international famine relief appeals by citizens prima facie provides stronger evidence of humanitarian bonds with strangers than does paying *compulsory* national taxes provide evidence of social ties with "the nation." Similarly participating directly or otherwise supporting international NGOs concerned with regional or global environmental protection is a palpable indication of bonds that transcend the nation, embracing a recognition of the importance of ecosystem integrity or personal integrity writ large. Accordingly there seems to be no good reason why the social steering systems of the administrative state cannot be directed in ways that serve *other* abstract communities as well as the national community, at least in circumstances where there is a recognizable community of concern around common problems or a palpable "community of the affected" that includes yet transcends the nation-state.

In any event, Habermas's major hesitation toward transboundary democratic will-formation turns on the practical and legitimacy requirements of law making, not on any moral argument concerning the appropriate scope of the political community, social solidarity and the public sphere. While democratic opinion- and will-formation may flourish best in communities with a shared language, culture and mass media, it can also take root in transnational communities that may also include different linguistic communities, particularly if the basis of connectedness—such as the suffering of bodily harm or ecological degradation—transcends language and cultural ties.

I have set up the foregoing interaction between Miller's civic republicanism grounded in belongingness, Habermas's communicative republicanism grounded in membership or legal association and Held's liberal cosmopolitan democracy grounded in affectedness in order to highlight the enduring tensions between otherwise intuitively appealing attempts to ground practical democratic procedures. From this encounter it could be argued that preexisting forms of social solidarity (which begin with embodied, face-to-face connections in the family and the local community) provide the preconditions for more extended forms of solidarity among strangers. But there ultimately seems to be no reason why this more extended form of solidarity should stop at the nation's borders, especially in view of the increasing interdependence and multinational character of national communities. In keeping with critical political ecology's method of exploring emancipatory possibilities within existing developments, the outstanding question should not be whether affectedness is a superior principle to membership as a basis for democratic citizenship. Rather the question should be how might the historical trajectory of extending citizenship *within* the nation-state also be pushed *beyond* the nation-state in ways that are both normatively defensible and practically feasible, bearing in mind that transnational democratic opinion- and will-formation must ultimately be given effect by means of law? In short, how might we practically move toward the affectedness principle embodied in ecological democracy as a framing structure of rule without annulling considerations of national and subnational belonging?

One way of approaching this challenge is to clarify the practical limitations of the affectedness principle. As foreshadowed, any attempt to operationalize the "all-affected" principle raises the question as to what criteria should determine the meaning and scope of affectedness (and whether it should apply to individuals and/or collectivities). This question does not detain civic republicans, since any member of the national community is free to participate on any matter of common concern arising in the national polity, regardless of whether they are affected in any material or personal way. It is enough that the matter be a matter of common or public *concern*.

Whereas mere membership of a national community is sufficient to ground and activate democratic rights for civic republicans, Held's particular working of the affectedness principle is derived from a more individualistic account of political life that presupposes arm's-length relationships among abstract carriers of rights operating under an overarching global legal system. Held's democracy is ultimately a liberal democracy of individuals, not communities.[33] In this respect Held's model is reminiscent of the traditional common law rules of legal standing for third-party actions, according to which a third party wishing to intervene in litigation needed to show that they were affected in some *particular* rather then merely general way (e.g., by showing particular damage to property, health, or pecuniary interest above and beyond others) before they were recognized by the courts. Merely "being concerned" about the matter as an ordinary member of the general public did not normally suffice. On this type of reading at least, if the affectedness principle were to serve as the *sole* basis for democratic participation within national communities it could be used as a basis for *restricting* participation only to those directly affected in relation to particular issues.[34] This would be unacceptable in terms of the enlarged thinking or role reversal that the processes of deliberation are supposed to generate (where the public testing of claims drives participants to approach political questions in terms of "what *one* should do," or "what *we* should do," rather than merely in term of "what is good for *me*"). On a strict and narrow application of the affectedness principle, Held's "citizens of the world" would not be able to seek legal redress or otherwise participate in the political or legal deliberations (concerning say biodiversity preservation) if they were not affected in any special or distinctive way over and above ordinary members of the public.

Now it must be pointed out that Held defends his multileveled framework of cosmopolitan democratic governance as *combining* the virtues of local, deliberative democracy with the necessity of abstract and highly meditated democracy to deal with global problems.[35] Moreover he applies his affectedness test to communities, not just individuals, in determining a priori what might be the appropriate level of governance in relation to particular issues. So although Held's cosmopolitan project is inspired and driven by the principle of affectedness, on the basis of a neo-Kantian defence of autonomy, he nonetheless also retains the principle of membership in relation to the various levels of governance that go to make up his multilayered, overlapping framework of global governance. Grounding a global or transboundary democratic order *exclusively* on the principle of affectedness is, it seems, impossible. In any event, it is in the nature of ecological problems and risks that they rarely affect only one person. Rather, they typically affect particular classes, communities or segments of communities rather than discrete individuals, in which case building a democratic order on the model of arm's-length relationships between abstract carriers of individual rights seems singularly inappropriate, as Held's practical reforms implicitly acknowledge.

It is this individualistic philosophical basis of Held's model that partly accounts for Habermas's (and my own) skepticism toward the project of building a global democratic public law. While Habermas supports the development of a restricted layer of global cosmopolitan law, at least as the basis for humanitarian intervention in support of gross human rights abuses, he considers the new global organizations outlined by Held would fail the requirements of legitimacy because they would not be able to pursue a "domestic politics" (which requires solidarity and redistribution).[36] That is, Held's cosmopolitan democracy is seen as mostly *reactive* and based on an abstract moral and legal order. Habermas, in contrast, argues that legitimate social steering arises from a democratic legal system that serves an existential, ethical-political community of citizens who are the authors and addressees of that law. A global polity would have no social boundaries between inside and outside, no *particular* democratic life and no meaningful sense of identity relative to other collectivities to enable any redistribution within the polity or indeed the extension of sympathetic solidarity to other collectivities. Further to this is Habermas's argument that democratic will-formation must be anchored in a delimited community of legal consociates residing in a delimited territory, applying the variety of forms of practical reason to produce mutually binding legal norms. Membership as a legal consociate also means that *all* citizens can join in the discussion on matters of common concern qua citizens, irrespective of whether they are individually affected by proposed norms in any material or particular way, since they will always be affected in a general way as *subjects* of the law.

To recapitulate the tensions, Held's ideal of cosmopolitan democracy glosses over the question of the social and linguistic bonds that might ground the kind of trust and reciprocal recognition that is required for meaningful participation and the enactment of legitimate law by legal consociates. Miller addresses these concerns, yet rests his defense of bounded national citizenship on a priori and ultimately arbitrary set of assumptions about the presumed benefits of membership in a *national* community as if it were the only community that engendered social solidarity (which is patently not the case). Habermas avoids the problems confronting Held and is more sensitive than Miller to the dangers of ethnic nationalism in his defence of constitutional patriotism over ethnic nationalism. Yet Habermas's "constitutional patriotism" may fail to ignite the same depth of human sentiment as, say, local attachment to place, or one's familiar linguistic or ethnic community. At the same time, it is unduly restrictive. If a mere commitment to democratic procedures is to serve as the basis of cultivating social solidarity in a multicultural nation-state, the door is also open to cultivating social solidarity in a multicultural world.

In the following section I seek to show that between the nostalgia for the democratic, relatively autonomous nation-state and an overarching global democratic law lies a third alternative that is different from Habermas's attempted reconciliation. I will seek to defend a more outward-looking transnational state that serves both the national community *as well as* other communities whose fate is tied in with decisions made by the national community. Such a state would offer something more than the thin atmosphere of Held's cosmopolitan democracy, but it would be more conducive than either traditional liberal or civic republican states to considering trans-species, transboundary, and intergenerational values and interests.

The argument that follows pursues the idea of a developmental progression of promising trends in ways that remain mindful of the insights of communitarians while also moving practically toward the ideals of cosmopolitans. Without knowledge of and attachment to particular persons or particular places and species, it is hard to understand how one might be moved to defend the interests of persons, places, and species in general. Local social and ecological attachments provide the basis for sympathetic solidarity with others; they are ontologically prior to any ethical and political struggle for universal environmental justice. Most environmental activists intuitively understand this and work from the premise of our unavoidable social and ecological embeddedness in particular places and communities. Yet it is impossible to arrest the growing gap between those who generate ecological problems and those who suffer the consequences, along with the increasing dis-embeddedness brought about by the processes of economic globalization, without developing sympathetic solidarity with environmental victims wherever they may be located. The transnationally oriented green state takes the next step and offers practical democratic procedures for ecological citizenship within and beyond the state.

7.3 The Transnational State as a Facilitator of Ecological Citizenship

While I have established that the principles of belongingness and affectedness are in tension, and that they both suffer limitations, it does not follow that they must be approached as mutually exclusive democratic options. Rather, it is quite possible to build upon and extend the communitarian principles of belongingness and membership with the cosmopolitan principle of affectedness in those cases where the broader transnational community is able to marshal and implement its democratic will by means of a common or mutually recognised set of legal systems. Moreover, working critically and reflexively with the historical trajectory or movement from membership to affectedness (and back again when the limitations of affectedness emerge) is not only consistent with the way in which citizenship rights have evolved historically but also more defensible in terms of critical political ecology's concern both to enlarge the spatial and temporal boundaries of the moral community and to intensify the focus on particular persons acting locally and responsibly in particular places. This inside-out-inside approach also avoids the two major dangers associated with Held's global outside-in approach to addressing transboundary political problems.

The first danger, as previously noted, is that the affectedness principle could be enlisted as a basis for *restricting* participation only to those directly facing the consequences of the proposed decisions or policies, or else conferring on those who are more affected than others a privileged say in political or administrative deliberations without the critical filter of deliberation with others who seek to be involved in matters of common (as distinct from particular) concern. In this sense the principle carries the potential to serve as a basis for exclusion rather than inclusion in political deliberations. That is, it would be a regressive move to insist that the democratic rights supposedly secured by the constitutional state could only be available to particularly "affected parties," since, as Saward points out, this would undermine the very concept of citizenship as an inclusive enduring achievement.[37] Yet as Saward goes on to argue, while the all-affected principle should not form the basis for deciding what should be the *primary* unit of governance, it can come into play as a qualifying structure of rule that compensates for the shortcomings of exclusive territoriality and citizenship.

The second danger associated with Held's project is that it is vulnerable to the charge of imperialism in that it is likely to further cement the hegemony of the dominant western states in the global order and undermine the principle of self-determination of nonwestern states and political communities.[38] As Andrew Linklater observes, cosmopolitanism has fallen on hard times in recent years in response to feminist, antifoundationalist, communitarian, and postcolonial critiques, all of which have pointed to the potential for domination that is inherent in all universalising perspectives, including liberalism.[39] From this perspective, cosmopolitanism threatens the survival of workable forms of community life by making traditional nonwestern and religious communities vulnerable to dissolution through rational criticism.

However, Linklater goes on to point out that, "seldom does the intellectual retreat from cosmopolitanism and the wider project of the Enlightenment lead to a celebration of ethnic particularism or patriotic loyalties which disavow all forms of answerability to others."[40] In this sense, a certain resonance with the Kantian principle of respect for others is still present in the work of those opposing cosmopolitanism. For example, inherent in the principle of respect for difference—or, in the case of Walzer, the right of self-determination of the tribes—is an appeal to universality, albeit a more minimalist morality. Taking his inspiration from Habermas, Linklater argues that we can avoid the unpalatable opposition between moral absolutes and incommensurable moralities by encouraging open and respectful dialogue with differently situated others. This, he argues, will enable a widening of the boundaries of community while also respecting difference. In short, he sees the discourse ethic as providing the most appropriate resolution to the tensions between communitarianism and cosmopolitanism. Linklater thus supports a kind of "thin cosmopolitanism" that seeks "the development of wider communities of discourse which make new articulations of universality and difference possible."[41] Yet while Linklater offers us an enticing understanding of cosmopolitan citizenship and the building of transnational public spheres, the role of the state (as distinct from the citizen) in his dialogic world order still remains shadowy and obscure.[42] In the remainder of this chapter, I suggest how the state might be brought into this picture, how it might facilitate ecological citizenship and how it might serve both national *and* transnational communities.

Rather than looking to global, overarching, and potentially imperialistic solutions to ecological problems, it is more in keeping with the method of critical political ecology to look for what Saward has called "intermediate and more fine-grained possibilities" that provide merely a qualifying or supplementary structure of rule to territorial governance.[43] What I am suggesting is that it is quite possible and feasible to transnationalize democracy in piecemeal, experimental, consensual, and domain-relative ways. Such an approach would enable the practical negotiation of principles in tension in response to particular transnational problems, rather than a priori. Formal democratic space-time coordinates would still need to come into play for the proper enactment of legal norms and for the substantive enjoyment of ecological citizenship rights in transboundary environmental domains, but these coordinates would not necessarily be the same for all domains (or for all problems within the same domains). Such a project would thus entail building upon, qualifying, and supplementing (rather than replacing) the principle of belongingness with the principle of affectedness. In this respect it is much more modest and feasible than Held's cosmopolitan democracy, and likely to be more achievable. Instead of permanent, nested layers of governance that require a "boundary court" to allocated issues to particular tiers in accordance with pre-determined principles, as suggested by Held, the most feasible means of extending democracy is by means of multilateral agreements between states that create overlapping, supplementary structures of rule that actively utilize existing territorial governance structures. Such an approach may be just as effective and certainly more straightforward (in the domains in which it operates) than Held's proposal. Such agreements, if successful, could then extend beyond ad hoc cooperative arrangements between particular governments to encompass a more enduring framework for cooperation. As I have argued throughout this inquiry, the point—after all—is not to replace states but rather to find more effective and more legitimate ways of addressing the shortcomings of exclusive territorial governance. As it happens, there are already several multilateral initiatives that have moved in this direction, which not only demonstrate the practical feasibility of such an approach but also point to the emergence of transnational states.

The most significant example is the Convention on Access to Information, Public Participation in Decision Making and Access to Justice in Environmental Matters (otherwise known as the Aarhus Convention) which was adopted in June 1998 and entered into force on October 30, 2001. This convention has been described by UN Secretary General Kofi Annan as "the most ambitious venture in the area of environmental democracy so far undertaken under the auspices of the United Nations."[44] Auspiced by the UN Economic Commission for Europe (ECE), the Aarhus Convention has already been signed by 40 countries and there are provisions for non-ECE counties to accede to the Convention. The Convention establishes rights to environmental information, public participation, and access to justice on the part of *all* citizens of those states who are parties to the Convention and in this sense it represents the first international convention dedicated to creating transboundary rights of ecological citizenship. It has also been singled out as providing the most explicit international legal recognition for not only procedural environmental rights but also a substantive human right to a healthy environment.[45] The environmental information provisions require member states to provide regular State of the Environment Reporting along with more active information on environmental policies and programs on the internet not only to its own citizens but also to citizens of member states. The Convention also provides a right of participation in environmental policy making, including the determination of environmental standards and the making of development decisions to all citizens of member states. Significantly article 2.5 provides that participation is open to "the public affected or likely to be affected by, or having an interest in, the environmental decision making; non-governmental organizations promoting environmental protection shall be deemed to have an interest." These provisions thus extend participation not only to those "expressly affected or interested" by the environmental matter but also to environmental NGOs who share a common concern about environmental matters.[46] While many environmental NGOs have argued that the Aarhus Convention does not go far enough, particularly in relation to its provisions on compliance,[47] it nonetheless must be regarded as a significant step toward transnationalizing ecological citizenship. The notion of shared due process rights, encompassing access to information, the courts, and the right to participate in technology and environmental impact assessments, has also been provided by the UN/ECE Convention on Environmental Impact Assessment in a Transboundary Context (adopted 1991, entered into force on September 10, 1997).

Multilateral treaties conferring citizenship rights of this kind are, of course, only one means of facilitating transboundary democracy. Other possibilities include deliberative forums, constituted as a microcosm of the transboundary community at risk, that facilitate deliberation and mutual understanding about matters of common concern. The representatives on such forums might be selected randomly from the general population of affected communities following the models of James Fishkin, or they might be formed by bringing together some or all of the existing elected representatives of affected communities (drawing on local, provincial, and national assemblies where appropriate).[48] Moreover the decisions emanating from such deliberative forums might range from the giving of purely advisory opinions to all governments in the region to the making of authoritative decisions in relations to particular issues (which may merely require ratification by national parliaments). Other possibilities, some of which are discussed by Saward, include reciprocal representation, and cross-border referenda.[49]

In all of these cases the cross-border arrangements would need to be authorized by a multilateral framework agreement negotiated by those states sharing ecological problems. While the political opinion-formation and, in some cases, will-formation would be transnational, in the end, the national or local parliaments and their administrative agencies would remain the final guarantors and implementers of the new transnational mechanisms.

Nonetheless, chapter 2 pointed out that the primary means of "unbundling territoriality" has been through multilateralism, and there is every reason to believe that this is the most likely way in which it will continue. As Saward argues, Held seems to have underestimated the difficulties in abandoning a single or base level of government and finding a relatively neutral and workable way of assigning jurisdiction on particular issues to different levels within his nested framework.[50] The upshot, as in federal systems, would be jurisdictional disputes (whether in the form of competition for jurisdiction—turf wars—or buck-passing) that would in all probability come to dominate relations between the layers of governance at the expense of debates about more substantive issues. However, once territorial state governance is conceded as the primary level, the all-affected principle becomes more attractive and may be institutionalized through a variety of creative procedures of the kind we have already mentioned.

Of course, it is also possible for states to enact unilateral measures to achieve the same or similar goals, and a number of such measures were explored in chapters 5 and 6. Further measures might include tribunes for noncitizens in which concerned local representatives could speak for the concerns of ordinary citizens of foreign states, "presenting their claims and responding to counter-claims of representatives of the host-state."[51] More generally, critical political ecologists would promote the cultivation of norms of nationhood and citizenship that are cosmopolitan, outward looking, and concerned not to displace problems to present or future "strangers"—a process that is most likely to emerge through the cultivation of a vibrant green public sphere. This could also allow the development of "environmental patriotism," building on the insight that attachment to local or national environments forms the basis of sympathetic solidarity with other communities and *their* environments. The upshot might be something like Daniel Deudney's "terrapolitan" (i.e., earth-centered) rather than Westphalian or cosmopolitan conception of sovereignty, communal identity, and legitimate political authority.[52] Such norms could also be built into the constitutional structure of the state. This might take the form of symbolic/aspirational statements of obligations to humankind and the global environment in state constitutions backed up by a range of new democratic procedures, due processes and decision rules that "open out" the moral and political community, including to the point where decision makers and the courts are obliged to consider the impact of their decisions on noncitizens in circumstances involving significant transboundary risks.[53] Nonetheless, these unilateral initiatives would be considerably buttressed by multilateral agreements that confer *reciprocal* environmental rights and obligations.

Taken together, these unilateral and multilateral initiatives would loosen the nexus between the nation and the state, while also redefining the more traditional understandings of nation such that the source of connectedness between those constituting any particular *demos* is no longer just belongingness or membership but also a common ecological embeddedness and a common capacity to suffer serious ecological or biological harm. Such risk communities may not necessarily be geographically contiguous or precisely specifiable in space or time. In the case of future generations, proxy forms of representation might need to be devised to ensure that future generations are systematically considered in any development or decision carrying major environmental consequences. I have also suggested the constitutional entrenchment of the precautionary principle as a particularly effective and parsimonious means of ensuring the systematic "representation" of future generations and nonhuman species. The *reciprocal observance* of this decision rule by all states would provide one powerful surrogate mechanism for institutionalizing ecological and social responsibility toward the relevant community at risk in those cases where that community transcends state jurisdictional boundaries.

In the foregoing institutional innovations the liberal/republican idea that the fundamental *source* of political authority lies in the the people would be maintained, but it would no longer necessarily always regard the people qua members of particular nation-states. Rather, the people would remain sovereign, but would be a more variable and fluid community made up of nations and all those who happen to belong, or who are likely to belong, to the relevant community at risk. In short, these new structures of democratic governance would be inclusive rather than exclusive of those outside the nation-state but within the ambit of the constitutional protection or multilateral agreement, at least in those circumstances where they may be seriously affected by proposed decisions. There is no reason why these structures cannot be grafted on to existing states—indeed, it is difficult to envisage how else they might be effectively and systematically institutionalised and legitimized. Such states would become what Ulrich Beck has called "transnational states," that is, states that have developed their sovereignty and identity beyond the national level.[54] However, these new norms and procedures would not replace the nation-state but rather would form an additional layer on top of preexisting rights of citizenship. In this sense the prefix "trans" in the adjective "transnational" should be understood to mean both *including* the nation *and going beyond* the nation, rather than displacing it. Here Habermas's broader argument about democratic procedures providing the basis for the cultivation of civic solidarity would also apply. That is, the efficacy of the green democratic state as a transnational state would depend in part on a commitment by citizens to the new environmental democratic procedures that the transnational state would uphold and foster.

Of course, this broadening of notions of citizenship, identity and community ought not to be confined to the green sphere. Andrew Linklater, for example, has pointed to a range of developments that are already providing an important challenge to exclusionary accounts of sovereignty and citizenship and going some way toward developing post-Westphalian communities made up of citizens who have multiple loyalties and diverse rights of political participation that are not confined to particular nation-states.[55] However, more than any other contemporary political problem, the ecological crisis invites a critical rethinking of the exclusive relationship of citizens and states to their territories in the same way that it invites a rethinking of the relationship of bourgeois individuals to their private property. Such a critical inquiry into territorial rule would strike at the state's "relationship to its territory, to a geographically distinct part of the globe, which constitutes the unique physical base and referent of the state's institutional mission, its very body, the ground of its being."[56]

7.4 Unit-Driven Transformation and the Power of Example

The most interesting point of contention in the debate about globalization and democracy is not whether transnational communities of public interest or transnational public spheres are desirable (indeed, it is difficult to argue that they are undesirable) but rather what institutions (state or suprastate) and what strategies might best facilitate, uphold and revitalize democracy. Yet these institutional and strategic differences should not be allowed to obscure what is common in the three approaches I have used as foils to the development of my alternative argument for the transnational green state. That is, all of the approaches discussed are, in varying degrees, concerned to explore how the democratic legal system might be made more effective as a social steering system vis-à-vis other steering systems, notably the economy. All are concerned to explore how powerful corporate actors, and economic decisions in general, might be made subservient to democratic politics. And all are concerned to counteract the depoliticisation of economic and technocratic decisions that generate ecological and social risks. Given the ubiquitous and transboundary character of so many social and ecological problems, the question as to how democratic politics might best catch up with markets is as central to Held's efforts to rehabilitate liberalism, Miller's efforts to rejuvenate republicanism and Habermas's efforts to revive social democracy in a liberal constitutional framework. My defense of the transnational green state is consonant with these broad goals, but stands somewhat between global cosmopolitanism, on the one hand, and traditional civic republicanism and "communicative republicanism," on the other hand. Strategically it may be understood as developmental (rather than just ad hoc and incremental) in the sense that it conscientiously seeks to extend promising domestic and multilateral trends in environmental governance that build upon the insights of communitarians while also moving practically towards the ideals of cosmopolitans. Such an approach necessarily proceeds by means of patchwork rather than comprehensive change.

Now it might be argued that the global reach of markets is such that only Held's cosmopolitan democracy can provide the kind of overarching democratic controls that would enable a democratic engagement between risk generators and victims. Shrinking state autonomy and the growth of common transboundary and global problems have necessitated the development of new, supranational democratic governance structures in the form of a global democratic public law that is available to all citizens of the world. Strategically this would mean working to change international rather than domestic structures of governance. Given the limited time available to avert ecological catastrophe, not to mention the environmental injustices wrought by globalization, merely working to improve democratic governance only at the domestic or regional level would appear too slow, uncertain, and haphazard. Paul Hirst's argument about the power of local example lends support to such an argument. According to Hirst:

> A world of examples cannot create an economic and political system. If the second best constantly fall behind the successful, missing every trick in forms of governance and economic performance, then in a world of localized experiments, some firms and areas will pull away from others and set quite different socio-economic standards. However, a patchwork world of this kind is unlikely to be a successful or sustainable one.[57]

In response to Hirst, I would say that the alternatives to the particular patchwork approach that I have advocated (which is a retreat into national communities or a heroic push toward global cosmopolitan law) are even less likely to be successful or sustainable in both social and ecological terms. Although Held's is the most concerted and comprehensive attempt to address the problems of globalization, it is not based on any strategic assessment of current developments or any analysis of the most propitious sites that might move the world toward his global cosmopolitan ideal. In fairness to Held, he does seek to ground his project in what he sees as the trajectory of the international system, but his reforms jump *too far* into a projected future to be likely to carry most states and peoples. Many states in the world are not democratic, and those that are democratic fall well short of his global welfare liberal ideal, which requires a massive domestic and international redistribution of wealth and income to secure democratic citizenship on a global scale. In this respect, however desirable this ideal in principle, he is asking all global citizens and states to run before many are able to walk or crawl while also denying the fact that many nonwestern political communities may not even wish to move in the direction of his liberal political idealism. In short, the preconditions for the establishment of Held's cosmopolitan democracy are simply not present. As Richard Bellamy and R. J. Barry Jones point out, on the supply side, alternative sources of power and intergovernmentalism are likely to thwart efforts toward a cosmopolitan democracy while, on the demand side, there is insufficient indication of the emergence of a shared global political culture. And so, "Without the preconditions of an absence of alternative sources of power, a sense of community and a suitable founding moment, global democracy is unlikely even when it appears both desirable and logically necessary."[58]

Moreover, as Alexander Wendt has pointed out, many national communities are likely to be resistant to the idea of cosmopolitan democracy on the ground that it involves a surrender of autonomy to de-centered networks of transnational power.[59] And Held's democracy of individuals leaves open the question of who decides who is affected by power and by what criteria.[60] But Wendt argues that it is not obvious that global democracy should be a democracy of individuals rather than communities, and that the philosophical unit of accountability should be the individual not the community. In any event, such arrangements would require the agreement of states, and they are unlikely to relinquish their juridical sovereignty. For Wendt, a better approach might be to work on the identities of citizens within particular states, to make them relatively more transnational in their orientation.[61]

The alternative I have defended takes this argument several steps further in seeking to build on the most promising multilateral developments that engage with existing democratic structures of like-minded nation-states, in ways that produce more outward-looking state governance structures (and not just outward-looking citizens).[62] Moreover such power as privileged western green democratic states might exert over other states should be by the "demonstration effect" (by the force of the better example or the better argument) not by diplomatic manipulation, blackmail, or conventional force. Chapter 2 emphasized that the international community is a multicultural community made up of many different modes and layers of relating between different states. Accordingly, while the movement from a Hobbesian culture of anarchy among states toward one based on mutual respect and open dialogue is always desirable, whether this takes the form of a pluralist, solidarist, or post-Westphalian community is a function of the histories and shared understandings of states. Those who happen to belong to and embrace an emergent green Kantian or green post-Westphalian community must recognize that other communities might not necessarily wish to become part of such a cooperative framework. To insist that they do, as Held does, would leave cosmopolitans open to the charge of green imperialism. Accordingly the peaceful and stable transition toward transnational green democratic states ought primarily to be a unit-driven transformation from below, rather than a system-imposed transformation from above that seeks to force recalcitrant units to adapt. To the extent to which a Hobbesian anarchy is transcended by the demonstration effect, the resulting international order would be variegated, made up of what might be clusters of transnational green states operating within a larger, less green and more traditional set of interstate relationships. This means that only where zones of affinity emerge among particular groupings of states—such as in the European Union—that a genuine transnational democracy becomes possible. However, it would not be global. To the extent to which such green clusters grow or are copied elsewhere, it ought to be by respectful persuasion or example (possibly hastened by the unwelcome assistance of ecological collapse). The proliferation of exemplary green states and communities of states does not threaten world order because they would not insist that other states remake themselves in their green image; other states and their societies should be free to appropriate those features that best fit with their own cultures and histories. Whether green states eventually proliferate to the point where they create a "critical mass" and change the character and practices of the society of states must remain an open question. But this concession to Hirst's argument does not make the alternatives singled out above any more attractive.

In chapters 2 and 3 it emerged that the credibility of those states that have taken leadership roles in multilateral environmental negotiations has rested in no small measure on their relatively more successful domestic environmental initiatives compared to other states. In short, the emerging picture is that the mutual democratisation of states and their societies appears to operate in a virtuous relationship with reflexive ecological modernisation at the domestic level and more active environmental multilateralism by such states (and their citizens) on the international stage. A proliferation of transnationally oriented green states, which are likely to extend and deepen environmental multilateralism, is also likely to provide a surer path to a greener world than the development of a more overarching cosmopolitan global democratic law.

8

Green Evolutions in Sovereignty

8.1 Green Evolutions in Sovereignty

Critical constructivists have emphasized the protean character of the norms associated with state sovereignty. Indeed, the old Westphalian ideal that supreme and final authority to rule should reside in territorially defined entities known as states, by itself, tells us very little about the shared understandings of what is legitimate state conduct in international society at the turn of the new millennium. As explained in chapter 2, this is because the ordering principle of sovereignty is a changing, derivative principle the meaning of which arises from the changing constitutive discourses that underpin it. Accordingly, to the extent to which the constitutive discourses of sovereignty have begun to absorb ecological arguments, it becomes possible to talk about the concomitant "greening of sovereignty."[1] The purpose of this chapter is to redeem the promissory notes issued in chapter 2 by drawing out some of the more significant shifts in understanding in the global discourses of environment, development, security, and intervention, to highlight the extent to which these new discourses have begun to shift shared understanding of legitimate state conduct in a greener direction, and how these developments might be furthered. Thus the concern is not merely to track these developments but also to draw out some potential emancipatory trajectories that might better promote environmental protection and environmental justice on a global scale.

The constitutional structure of international society refers to the fundamental norms that define political authority. These norms specify who may exercise political authority and the circumstances when it may be legitimately exercised. In the context of the modern system of sovereignty states, Daniel Philpott has encapsulated these fundamental norms into what he calls the "three faces of sovereignty": the holders or recognized units of sovereignty, the standards of membership (in terms of who may be admitted), and the prerogatives of sovereignty.[2] He has argued that whenever there is a change in *any one* of these three aspects of sovereignty, it is possible to talk of a "revolution in sovereignty."[3]

As Philpott explains, the Peace of Westphalia in 1648 "christened modern international relations" by parceling out political authority to territorially defined units, known as states, which were understood to have supreme authority to enact laws in relation to all people within the territory (including citizens, residents, and those who are merely passing through) and immunity from outside interference.[4] Such a constitution rendered international society anarchical, in the sense that states were not subject to any higher authority. In specifying the external dimensions of sovereignty, Westphalia left it open as to how sovereign rule might be practiced internally (e.g., it could be exercised by means of a democracy, theocracy, or military dictatorship).

However, as J. Samuel Barkun and Bruce Cronin have shown, the idea of the externally recognized legitimacy of the territorially defined state to rule within its borders was to come into tension with the emergence of nationalism and the idea of the nation, which developed after the juridical understanding of the territorial sovereign state had emerged.[5] For nationalists, legitimation of rule stems not from territorial boundaries but rather from the bonds that held together particular communities, defined by particular linguistic, cultural, and social ties. Whereas *state* sovereignty emphasizes the integrity of territorial *borders*, *national* sovereignty emphasizes the link between authority and *peoples*. Nonetheless, the international norm since World War II has been a cosmopolitan norm of civic nationalism rather than ethnic nationalism. At the same time many states are now making greater concessions toward ethnic minorities and indigenous peoples *within* their own borders (e.g., Canada, Belgium, and Spain), although in these cases the acknowledgment of the right of some measure of self-determination does not extend to the granting of statehood. In the previous chapter it was suggested that, in the ecological domain at least, it is possible to glimpse the emergence of transnational states that serve more than one national community both within and beyond their territorial borders. As constructivists emphasize, these new norms of recognition—including the recognition of the environmental rights of so-called foreigners—cannot be explained simply by reference to power politics or instrumental calculation. Moreover they underscore how far shared understandings about legitimate authority to rule have traveled since the Peace of Westphalia.

The standards of membership of the society of states, while originally Eurocentric in orientation, have also changed. The colonial revolution in the post–World War II period brought to an end the old European "standard of civilization" test of membership and made it possible for former colonies to claim a right to political independence irrespective of their level of development, culture, institutional capacity, or internal political control. The emergence of postcolonial states was, as Philpott shows, a significant revolution in sovereignty that also extended the state system to virtually all parts of the globe. However, one consequence of this postcolonial proliferation of what Robert Jackson has called quasi-states has been the emergence of a normative discourse of negative sovereignty (freedom from outside interference) over positive sovereignty (effective, internal rule).[6] This discourse of negative sovereignty stands in uneasy tension with the discourse of democratic rule, championed by a growing array of state and nonstate actors. Nonetheless, democratic norms are widely considered part of good governance, as exemplified in the recent practice of multilateral lending agencies to impose democratic conditionality on developing countries in return for debt relief, investment funds and assistance in capacity building. To the extent to which the international society—through its multilateral agencies—increasingly expects internal rule to be democratic before financial, technical or development assistance is forthcoming, it is also possible to talk of the emergence of a positive sovereignty discourse that requires a democratic standard of membership. The previous chapter did not go so far as to insist on a *green* democratic standard of membership as a condition of multilateral assistance to states in need, since it was concluded that voluntary emulation and consensual assistance was better than stipulation in this particular context. Nonetheless, it is not inconceivable that something like a green democratic standard of membership might emerge over time by emulation, particularly if it emerges that green democratic states are able to provide a higher overall quality of life to their citizens relative to other states.

Finally, the prerogatives of sovereign states, that is, the rights and responsibilities that attach to the recognized sovereign units, have also changed over time, and these changes have often accompanied changes in *other* faces of sovereignty. For example, Christian Reus-Smit has shown how international human rights norms provided the *moral foundation* for decolonization and the construction of the new sovereignty regime after the Second World War. That is, the first wave of postcolonial states succeeded in artfully linking the rights of self-determination and development (which were basic to the new sovereignty regime) to the satisfaction of social and economic human rights.[7] More recently developing countries have enlisted the sustainable development discourse in ways that serve the social and economic human rights of their peoples, rather than the rights of nonhuman nature.[8] In contrast, environmental NGOs (e.g., Greenpeace), intergovernmental organizations (e.g., IUCN), and some developed countries have linked sustainable development to an emergent discourse on environmental rights, which extends to include nonhuman nature.[9]

Significantly there has been a considerable growth in the environmental responsibilities of states. The traditional rights and responsibilities of states were confined to such matters as the right to establish embassies, the right to enter into treaties, the duty to observe treaty commitments and the responsibility to observe the rules of war (including the duty not to invade the territories of other states). More recently, particularly in the wake of the postcolonial revolution, developing countries have added to this list the right to develop. However, the new international discourse of sustainable development seeks to qualify the traditional prerogative of states to develop their natural resources as they see fit. At the same time the emergence of the doctrine of humanitarian intervention to protect against gross human rights violations represents a more recent revolution in sovereignty that directly challenges the more traditional principles of nonintervention. Some of these more recent challenges to traditional sovereignty discourses have come from nonstate actors, notably, transnational advocacy networks promoting human rights and environmental protection.[10]

These shifting norms (concerning the holders of sovereignty, membership standards, and state prerogatives) are highly significant in any exploration of the greening of sovereignty, not the least because they point to significant tensions in the shared understandings of the international community. Consistent with critical political ecology's method of immanent critique, my concern is to explore to what extent these tensions might contain openings for the development of greener discourses of sovereignty. To the extent to which it is possible to track, and find ways of enhancing, ecological revisions to *any* of the three faces of sovereignty—to the bearers of sovereignty, the standards of membership, and the prerogatives of statehood—one can expect to see further "green" revolutions in sovereignty, to adapt Philpott's analysis. Yet it is also possible to have green *evolutions* in sovereignty, that is, gradual changes that are not cataclysmic and that do not appear to amount to much in the short term. Such evolutions in sovereignty may nonetheless amount to significant change when viewed from a longer term perspective—say over a period of fifty to one hundred years rather than just ten or twenty years. As it turns out, it is evolutionary change of this kind to the *third* face of sovereignty (i.e., the rights and responsibilities of states) where the most significant ecological developments have taken place, and this is where the discussion will be directed. This is also the place where reconstructive efforts to extend the greening of sovereignty are likely to be most fruitful.

In teasing out these developments, it is useful to explore the different levels at which environmental discourses can work up and down (from domestic to international and back again) to transform understandings of the rights and responsibilities of states. It has already been noted how a focus on shifts in shared understandings in the constitutive discourses of sovereignty reveal sovereignty to be ultimately a derivative value. However, these shifts rarely occur on only one level but rather build up from a number of different sources and levels, and are promoted by a range of different actors. Christian Reus-Smit has identified four levels at which normative transformation can take place:[11]

1. To issue specific regimes
2. To the institutions of international law and diplomacy
3. To the organizing principle of sovereignty
4. To the underlying purpose of the modern state

Applying these four levels to developments in international environmental law and policy, I show in the following section how changes mostly in the first level of analysis have the potential to work back on the meaning and operation of the organizing principle of sovereignty. That is, the issue specific environmental regimes, declarations, and strategies (level 1) that have proliferated as part of the evolving processes of environmental multilateralism (level 2) have gradually redefined the rights and responsibilities that are attached to sovereign statehood (level 3), which have produced new discourses on the underlying purpose of the state (level 4). Although the developments at levels 2, 3, and 4 are merely emergent (and highly contested), they nonetheless represent a significant challenge to dominant understandings of multilateralism, the sovereign right to develop and the role and purpose of states.

A similar analysis may be applied at the domestic level. In chapter 3, I singled out four different levels at which change can be detected at the domestic level. These levels comprise:[12]

- Change in policy instruments
- Change in policy goals
- Change in policy paradigm, or the hierarchy of policy goals
- Change in the role of the state

I showed how weak ecological modernization entailed changes only at the first two levels, whereas the critical questioning associated with strong ecological modernization propelled discussion toward the hierarchy of goals behind environmental and development policy settings, along with the purpose and role of the state. Understanding and linking these developments at the domestic and international levels serves to blur the traditional boundary between domestic and international affairs, and underscores the mutually constitutive interactions between these two realms of social action (which are explored further below).

To the extent to which the constitutive norms of sovereignty have absorbed these new discourses, one might also say that sovereignty, like the processes of modernization, has become reflexive in adapting to global environmental change. In both cases these new discourses carry intimations of a deeper change in the roles and fundamental rationale of the modern states. If, as Reus-Smit argues, the underlying purpose or rationale of the modern liberal state has been to uphold the security, welfare, and ultimately the autonomy of its citizens by facilitating the processes of economic modernization, then our question is: To what extent have these processes been ecologically realigned?[13] In tracking some of the major shifts in shared understandings and practices in global society since the emergence of modern environmentalism, which is typically dated from around the early 1960s, I will point to intimations of a shift in the purpose of the modern state from environmental exploiter and territorial defender to that of environmental protector, trustee, or public custodian of the planetary commons. Of course, we have by no means arrived at this point, and there is no inevitability about this possible green trajectory. Nonetheless, this is a *potential* trajectory that critical political ecologists and green activists ought to find ways of "summoning" and fashioning out of the multiple discourses identified above.[14] Moreover the normative materials are already at hand, in the sense that there are plenty of analogies and precedents in the existing stock of international normative discourses for making this move (particularly in the discourses of intervention and nonintervention). Of course, there is also a range of normative tensions and political fault lines that may well block such developments, in which case part of the task of critical political ecology is also to find ways of resolving these tensions to the satisfaction of environmentalists, social justice advocates and developing countries.

Before undertaking this task, it is necessary to make mention of another dimension of the contemporary sovereignty debate, and that is the question of the *practical* political autonomy and institutional capacity of states to promote ecologically responsible practices at home and abroad—questions that were explored in depth chapter 3. Karen Liftin, for example, has argued that, for those of us who are interested in the greening of sovereignty, the most important place to look is not juridical sovereignty but rather operational or "effective" sovereignty. Accordingly she has developed a more pragmatic framework for analyzing the greening of sovereignty based on an examination of shifts in actual (as distinct from juridical) autonomy, control (or capacity) and state authority. This, she argues, directs attention to the actual external and internal practices of states rather than to legal formalism.[15] Similarly Stephen Krasner has contended that if one looks to the actual practices of states, then the realist principle of self-help is more relevant than, and contradicts, the principles of self-determination and nonintervention that make up the principle of sovereignty.[16]

However, the focus of this chapter is confined to the question of the changing basis of legitimacy of the international order, to the *recognized norms* that define rightful and wrongful state conduct. Now not all such norms are observed or enforced in practice, as Krasner points out. However, where they continue to provide the basis for determining rightful conduct and censuring wrongful conduct, they may be said to remain *recognized* norms. For example, the norms of just war and the international norm against torture are often breached by states, but these breaches do not thereby serve to annul the respective norms if such breaches are always widely condemned. Of course, widespread systematic breaches of norms by large numbers of states can eventually lead to the annulment of norms, particularly when there is no or only perfunctory censure by the community of states. However, from the point of view of legitimacy, the important question is always whether the norm continues to function as a commonly recognized yardstick for evaluating rightful or wrongful conduct.

Norms of rightful conduct may be found in international law (treaties, customary law) but not necessarily confined to formal law. For example, they may be found in international resolutions, declarations, strategies, and action plans. In this respect, unlike international lawyers, critical constructivists are less concerned to search for a hard and fast line between soft or hard law, or the prelegal and the legal, since these divisions can still beg the question of legitimacy. That is, so-called soft law declarations may prove to be significant if they are taken seriously in terms of practice and/or moral censure. Ultimately it is not the source or process of rule-formation but rather what the international community recognizes as legitimate norms that is our concern. Thus it can be seen that some norms can be treated with greater seriousness than others even though they have the same formal legal status as, say, treaty norms. For example, although all human rights are declared indivisible, the breach of some human rights is designated an "international crime against humanity" and the subject of investigation by the International Criminal Court.

From a critical constructivist perspective, then, exploring shared understandings (and how they are reached, maintained, and/or transformed) is essential to understanding both continuity and change in any social structure. Changes in shared understandings about the meaning and purpose of social life can undermine and transform the preexisting basis of the legitimacy of particular social structures, often by working at their constitutive principles. Thus the emergence of new international norms may not only change the way states and other social agents are coordinated and regulated, they may also serve to re-constitute the interests and identities of social actors by re-defining the set of practices that make up social activity, by revising the purpose of such activity, and by reconfiguring who is a legitimate actor and what is legitimate conduct in particular social contexts.

8.2 New Developments in Global Environmental Law and Policy

8.2.1 Environmental Multilateralism: General Developments

While it is not possible here to provide anything like a detailed account of the hundreds of environmental multilateral treaties, declarations, and strategies that have been agreed among states over the past three to four decades, it is nonetheless possible to point to a number of significant developments in terms of norms, rules, and structuring principles of multilateral environmental governance that together cast a new ecological light on some of the basic discourses that have given sovereignty its meaning in the immediate aftermath of World War II.

The striking character of many of these changes can be highlighted when it is remembered that the quality of the global environment was not a concern of the drafters of the United Nations Charter. Rather, in the wake of a second major World War and the shocking legacy of nazism, the two principal concerns embedded in the United Nations Charter were to uphold international peace and security and reaffirm faith in fundamental human rights. Nowhere does the Charter mention the environment and to the extent to which the United Nations has subsequently become involved in environmental governance, it has been through a broader interpretation of its responsibilities to promote *social* progress (primarily through the work of the Economic and Social Council). More recently the overriding security objective has also been given an ecological reinterpretation (explored below). The period immediately following the establishment of the United Nations was dominated by cold war politics and the processes of decolonization, which included a major focus on the rights of self-determination and development of the new states admitted to the UN system. These developments provide the broader context in which environmental multilateralism began to emerge.

To the extent to which multilateral environmental initiatives were undertaken in the 1950s and 1960s, they were largely ad hoc and uncoordinated. It was not until the 1972 United Nations Conference on the Human Environment (UNCHE)—the Stockholm conference—that environmental concerns became a prominent matter of international concern and a new responsibility of the United Nations. The Stockholm Conference gave rise to a normative statement of global principles (the Stockholm Declaration on the Human Environment), a new UN organization (the United Nations Environment Program, UNEP), a program of action and an Environment Fund to facilitate implementation. A similar suite of measures emanated from the second major international environment conference held twenty years later at Rio de Janeiro in 1992 (the United Nations Conference on Environment and Development, UNCED)—measures that were reaffirmed at the World Summit on Sustainable Development ("Rio plus 10") held in Johannesburg in 2002. These measures comprised a statement of principles (the Rio Declaration on Environment and Development), a new UN organization (the Commission for Sustainable Development), an action plan (Agenda 21), and new funding facility (the Global Environment Facility). While the Stockholm Declaration is widely understood as laying the foundations of international environmental law, the Rio Declaration now serves as the most significant and up-to-date normative touchstone for understanding new developments in the field.

During this period hundreds of bilateral and multilateral environmental treaties have been concluded. According to Edith Brown Weiss from the early 1970s to the early 1990s, the international community produced nearly nine hundred international legal instruments that are either primarily directed towards environmental protection or contain important provisions relating to environmental protection.[17] Moreover these instruments have expanded and evolved in terms of subject matter, scope, and design. For example, alongside agreements on transboundary ecological problems there has grown a new raft of agreements dealing with genuinely global problems (e.g., ozone depletion, climate change, and the erosion of global biodiversity). There has also been a discernible shift in the scope of regulation, which has shifted from end-of-pipe regulation to prevention at source, and from the protection of individual species to the protection of entire ecosystems. The speed at which the international community has negotiated complicated and controversial agreements has also increased (at least by international standards), with many new agreements reaching conclusion within the space of two years, although it usually takes another three or more years before the agreements come into force following ratification.[18] Such relative efficiency compared to earlier periods may be attributed not merely to technological developments in transport and communication but also to the adoption of the popular Framework Convention/Protocol model, which has now largely superseded the fully codified convention of the kind pursued in the Law of the Sea Convention (which took nine long years to conclude). The device of the Framework Convention has enabled countries to reach consensus on broad principles and goals, which are subsequently fleshed out by more detailed and binding protocols settled by ongoing conferences of the parties.

The typical Framework Convention approach has in some ways forced a more open principled approach to address environmental problems. This is precisely because greater attention must be given in the preliminary phase of negotiations to laying out broad goals and norms before specific details are addressed. However, Framework Conventions have also attracted criticism for their excessive generality and "normative vagueness" in that specific regulation is delayed, making it difficult to ascribe rights and responsibilities to particular states for implementation and enforcement. Yet such agreements have also become more sophisticated. They now include a large array of positive incentives to poorer countries to comply, and to non-parties to join, in the form of technology transfer, funding, and assistance in building national capacity to deal with environmental problems.

Despite the ongoing problems associated with implementation, monitoring, and enforcement, and growing congestion of treaties, it is nonetheless possible to observe cross-references among treaties, declarations, and strategies and a degree of convergence in many of the common principles embodied in conventions and declarations, many of which were brought together in the 1992 Rio Declaration. Chief among these developments are the following:

- New principles of world heritage, common heritage, and global heritage
- The principle of sustainable development, which incorporates the principle of intra- and intergenerational equity (the core component of the Brundtland formula for sustainable development)
- The principle of common but differentiated responsibilities, which is related to the previous principle, and acknowledges the different capacities and abilities of developed and developing countries to respond to global environmental change and pursue sustainable development strategies
- New principles for dealing with environmental risk, scientific uncertainty, and liability (the precautionary principle and the polluter pays principle)

Alongside these new principles there is also the emergence of a new discourse of environmental rights as procedural rights as well as human rights.[19]

Moreover many of these principles have not only formed the basis for the subsequent development of a wide range of environmental treaties, but they have also found their way into constitutions, domestic laws, and sustainable development strategies of various states.[20] Together they represent an attempt to qualify the traditional development prerogatives of states.

These major international developments must also be understood in the context of new domestic developments. The late 1960s and early 1970s also marked the beginning of a period of significant innovation in domestic environmental regulation, with the US National Environmental Policy Act of 1969 (which set up the US Council on Environmental Quality, the Environmental Protection Agency, and the procedure of environmental impact assessment) serving as an international model that has been directly copied in numerous jurisdictions without the mediation of international institutions. In other cases practices pioneered in the domestic realm in some countries have been adopted by multilateral lending agencies, and then imposed as a condition of lending in other countries. For example, the World Bank now requires both environmental and social impact assessment of projects attracting bank funds. Ongoing cross-fertilization between national and international environmental law and policy has also been a prominent feature in recent decades. To cite one particularly exotic example of intercultural cross-fertilization of environmental norms: the Supreme Court of Pakistan has recently incorporated the precautionary principle contained in the 1992 Rio Declaration into the constitutional jurisprudence of the Islamic Republic of Pakistan.[21] The precautionary principle had originated in the domestic law of the Federal Republic of Germany during the 1980s during a period of rising public concern about ecological risks.[22] Thus domestic policy experiments in environmental impact assessment and decision rules concerning risk can be seen to have influenced the development of international environmental policy, which has, in turn, influenced domestic policy in states that might not have otherwise engaged in such environmental policy innovation in the absence of new developments in environmental multilateralism.

Throughout this period of environmental change, states have by no means been the only instigators, authors, subjects, and enforcers of international environmental law and policy. Rather, international environmental law, particularly the processes of environmental treaty making and formulating declarations, has become a major arena for discursive battles about the future shape of international society for both state and nonstate actors. Moreover international law is increasingly used to regulate the conduct of, and protect the integrity of, not only states and their territories but also nonstate actors and nonstate territory. Thus the range of authors and the addressees of international law have expanded. Nonstate actors, ranging from environmental nongovernment organizations (NGOs)—corporations, the media, scientists, policy think tanks—to international organizations have played a crucial role in identifying and publicizing ecological problems, developing policy relevant knowledge, research and political agenda setting, negotiating policies and rules (sometimes as members of official delegations), monitoring, and implementation.[23] At the Kyoto climate change negotiations in December 1997, for example, states were considerably outnumbered by nonstate actors, who made full use of modern communications technology such as the Internet and mobile phones to stay in touch with relevant delegates and negotiating texts in the formal conference and the reaction of constituents elsewhere in the world. Indeed, official sessions and selected events at the negotiations are now broadcast live not only to TV screens in the official conference building but also worldwide on the Internet, creating high visibility and high expectations. From the perspective of the formal negotiators, the world has been—literally—watching, a fact that has tended to encourage compromise and deter highly obstructionist tactics.[24] (At the same time the use of such technologies put many state and nonstate delegates from developing countries at a considerable disadvantage.) Environmental NGOs and the media, in particular, have served as a "critical court of appeal" and watchdog in the processes of treaty making and implementation, amplifying that all important element of publicity to environmental negotiations.[25]

While on the international stage the formal lines of accountability and responsibility between the officially recognized treaty negotiators and global civil society remain weak and ill-defined, these lines nonetheless operate in informal and diffuse ways. Thus states still remain the decisive political actors in terms of negotiating the formal texts of treaties, particularly during the final stages of negotiation. Yet these last minute negotiations must always be set against a long and complex background of highly contested domestic and international problem definition, agenda setting and policy debate in which nonstate actors such as environmental NGOs, scientists, and corporations have played a major role.[26] Indeed, the significance of the contribution of NGOs and scientists is now given formal, albeit sporadic and limited, recognition in formal treaty texts.[27] The role of NGOs and civil society is therefore, important to any understanding of the processes of *legitimation* in multilateral negotiations. The future legitimacy and effectiveness of treaties may well increasingly depend on further formal recognition of nonstate actors at all stages of the treaty-making process.

Applying the four-level analysis of change introduced above, it can be reasonably concluded that there have been significant changes in levels 1 and 2. That is, there has been a significant shift in the principles, subject matter, scope, and sophistication of environmental treaties and declarations, and environmental multilateralism has evolved to include (albeit informally) nonstate actors to a much greater degree than in the past. Let us now explore to what extent these new norms and practices have given rise to green evolutions in sovereignty and changes in the roles and purposes of states.

8.2.2 State Responsibility for Environmental Harm

While there are many reasons to be cynical of the foregoing developments, particularly when set against the practical environmental outcomes of environmental treaty negotiations, it must be emphasized that enormous challenges have been confronted resulting in significant ecological revisions to global norms—especially when set against the longer history of international law. In the time before ecological problems had assumed the character of a global crisis, public international law had sought to reconcile the internal and external dimensions of state sovereignty (i.e., self-determination and noninterference) according to the customary international law principle *sic utere tuo ut alienum non laedas* (use your own property so as not to injure that of another). Such a formulation approached the problem of transboundary environmental harm as one of regulating the use of territorial rights, rather than protecting victims or ecosystems per se. In effect, the basic customary law principle for state responsibility for transboundary pollution provides that no state may use its territory in ways that cause serious injury to the territory, property, or population of another state.[28] The scope of state responsibility was akin to the common law tort of nuisance and extended to activities within the territory and/or control of the state, including unauthorized activities.[29]

The doctrine of state responsibility for environmental harm is fraught with uncertainties, and it has proved to be both reactive and limited in its application.[30] The victim states usually have to wait for some measurable degree of environmental damage to be done before compensation can be sought, as the injury has to be tangible, serious, and calculable in monetary terms (this essentially means property damage, financial loss or physical injury to citizens and does not extend to psychological damage, moral concern for future generations or damage to wildlife, ecosystem integrity or aesthetic quality), and the victim state is obliged to prove causation in clear and convincing terms while also establishing lack of due diligence on the part of the defending state.[31] Nor could damage to areas lying outside state jurisdiction, including the global commons, be actionable if no *state* could be shown to suffer harm. Merely proving a more or less direct chain of causation between activities in the offending state, and damage to territory or persons in the victim state is not enough, by itself, to found a claim. The victim state must also prove that the activity within the offending state is *unreasonable* in the circumstances, that it is in dereliction of some duty vis-à-vis the victim state. The test of due diligence has been defined in somewhat circular terms as the "diligence to be expected from a 'good government,' i.e., from a government mindful of its international obligations."[32] So, for example, if the offending state can show that it applied "all necessary and appropriate measures" or "best practical standards" to prevent pollution or other environmental damage, the victim state would have difficulty in showing unreasonableness even though it may have suffered considerable damage caused by the defendant state. While intersubjective understandings of what is "reasonable use" of a state's territory have changed over time, the presumption embedded in this rule continues to favor the *use* of territory over the *protection* of territory.

The doctrine of state responsibility for environmental harm provides very limited scope for a potential victim state to prohibit another state from conducting activity within its territory that might pose a *risk of damage* to the victim state. The case brought before the International Court of Justice by Australia and New Zealand against atmospheric nuclear testing in the Pacific by France underscored the limitations of this customary law principle. In this case it was admitted by Judge Ignacio-Pinto that while reparations could be required for actual damage suffered, there was no means whereby states that were anxious about *potential* nuclear fallout could legally prohibit another state from using its territory in ways that exposed them to the future likelihood of nuclear fallout.[33] In this respect the doctrine provides very limited scope for heading off or turning around complex global environmental problems such as atmospheric pollution (including global warming), loss of global biodiversity, and pollution of the high seas. There is nothing preemptive about this particular doctrine, nothing that might encourage or require states to take anticipatory measures (beyond prevailing practice) to avoid possible serious or irreversible harm.

Clearly, then, the presumption in favor of the right of states to use or develop their territories as they see fit has cast a very heavy and arguably very unfair burden on victims. There are striking parallels here with developments within domestic tort law in most western nations. The common law torts of negligence and nuisance were based on a presumption in favor of the freedom of property holders to use their property as they see fit, leaving it to victims to prove a duty of care, damage, and causation.

Just as the common law rules regulating the rights of property holders proved to be inadequate in dealing with the growing scale and gravity of ecological problems and have increasingly been overshadowed by new environmental legislation, so too the existing customary rules of international law proved inadequate and have been increasingly superseded by treaty law. Not only was the case by case approach too slow, piecemeal, and conservative to respond to novel and pressing problems and situations, the basic presumptions in favor of property holders/territorial states has meant that both the common law and customary law have put environmental victims at a considerable disadvantage vis-à-vis environmental polluters and despoilers. In short, the proliferation of domestic environmental legislation and multilateral environmental treaties and declaratory instruments since the 1970s may be seen, among other things, as compensating for the shortcomings of the common law and customary law. Both the domestic legislation and the multilateral instruments have gradually redefined the rights and responsibilities of states in a greener direction. Moreover these innovations have increasingly taken place in more discursive policy-making settings including a wider variety of state and nonstate actors, as well as international environmental NGOs.

Yet while the proliferation of treaties and declarations has overcome some of the obvious limitations of customary international law, some of the ambiguities have persisted insofar as the basic principle of state responsibility for environmental harm has been carried forward as part of this new wave of norm development. For example, the so-called Trail Smelter principle has been carried forward, with some modification, in the principle 21 of the 1972 Stockholm Declaration, which provides as follows:

> States have, in accordance with the Charter of the United Nations and the principles of international law, the sovereign right to exploit their own resources pursuant to their own environmental policies and the responsibility to ensure that activities within their jurisdiction or control do not cause damage to the environment of other States or of areas beyond the limits of national jurisdiction.

Principle 21 of the Stockholm Declaration expands on the Trail Smelter principle by including state actions "within their jurisdiction or control" rather than just actions within their territory while also not qualifying "damage" to "serious damage." Moreover principle 21 extends the scope of protected areas by including "areas beyond the limits of national jurisdiction." This includes not only the territories of other states but also areas outside the jurisdiction of any state, such as areas of common heritage (although exactly who might be recognized as having standing to bring such actions is an open question). Significantly, however, principle 21 also articulates states' responsibility for environmental harm in the context of the principle of permanent sovereignty over natural resources, which itself is regarded as "a basic constituent of the right to self-determination" and therefore an essential element of state sovereignty.[34]

While there is ongoing dispute among legal scholars as to whether principle 21 represented *opinio juris,* it has nonetheless subsequently been endorsed by the UN General Assembly as laying down the basic responsibility of states in relation to the environment.[35] Principle 21 has also been reiterated in a wide range of subsequent declarations and treaties,[36] including principle 2 of the Rio Declaration, albeit with the modification, that "states have . . . the sovereign right to exploit their own resources pursuant to their own environmental *and development* policies. . . ."[37] Some commentators have argued that this aspect of the Rio Declaration actually represents a backpeddling from the Stockholm Declaration, since greater prominence was given to development concerns by the conference delegates in 1992 compared to 1972.[38] Taken together, however, the two declarations mark a significant step forward from the status quo prior to 1972. Nonetheless, the tensions in principle 2 of the Rio Declaration continue to provide a window into a more fundamental set of disagreements concerning the environment and development priorities of the developed and developing world—disagreements that are reflected in competing sovereignty discourses.[39]

8.2.3 The Right to Develop: Economic versus Environmental Justice?

At the time of Stockholm, environmental and development matters were largely treated separately, whereas by the time of the Rio summit, they had been brought together in the umbrella concept of sustainable development. However, the emphasis placed on sovereign rights over natural resources in principle 21 of the Stockholm Declaration must be understood in the more immediate postcolonial context in which it was declared. In the 1950s and 1960s the processes of decolonization, self-determination, and development had became paramount concerns of new states, many of which were concerned to wrest control of assets formerly owned and exploited by colonial powers. An important facet of the right to develop was the right of permanent sovereignty over natural resources, which had been championed by newly independent states as providing the means for advancing the welfare of their people. As early as 1952 the General Assembly of the United Nations recognized that "the right of peoples to use and exploit their natural wealth and resources is inherent in their sovereignty."[40] This right came to full fruition in the 1962 Resolution on Permanent Sovereignty over Natural Resources, which declares that "The rights of peoples and nations to permanent sovereignty over their natural wealth and resources must be exercised in the interest of their national development and of the well-being of the People of the State concerned. . . ."[41] The resolution made no mention of the environment since this was not the primary purpose of the resolution. The principle reappeared in article 4(c), The Establishment of a New International Economic Order (NIEO),[42] and article 2, The Charter of Economic Rights and Duties of States.[43] It is noteworthy that the principle had originally been formulated as a *human right* belonging to peoples or nations that had been subjected to colonial rule to freely dispose of their natural wealth; since Stockholm, however, it is more typically formulated as a right belonging to sovereign *states*.[44]

While the principle of permanent sovereignty over natural resources, and the more general "right to develop" (which was given expression in the 1986 UN Declaration on the Right to Development), is today widely regarded by environmentalists as standing in the way of enlightened global environmental governance, it is nonetheless necessary to emphasize that it was not originally formulated in an environmental context. Rather, it played an essential role in the new sovereignty game by providing the normative basis for newly independent states to restructure onerous concessions made to foreign corporations during the colonial period. In particular, it facilitated the development of more equitable legal arrangements for natural resources development (particularly petroleum development) in the late 1950s and 1960s that improved the financial returns to the host state, at least relative to the former colonial period.[45] Against this background, it is therefore not surprising that many developing countries should subsequently regard the new wave of international environmental norms and the green conditions imposed by multilateral financial institutions in more recent decades as yet another expression of Western imperialism. From a developing country perspective, the wave of international environmental regulation has interrupted an all too brief honeymoon of relative autonomy from the clutches of colonial rule.

As Adil Najam has explained, developing countries believed they came to the Earth Summit in 1992 "with morality on their side" (in view of the legacies of colonialism manifested in the structural inequities in the world trading and financial systems) but, paradoxically, left the Summit "appearing as pariahs, resistant to 'environmental progress.'"[46] In seeking to reiterate the goals of the New International Economic Order at that time, developing countries pointed to the many ways in which the present structure of the international economic system constrained their very capacity to participate in international cooperative efforts to protect the environment. From the perspective of the South, then, the right to develop (autonomously and sustainably according to their own cultural frameworks and environment and development priorities) was defended in no less *moralistic* terms than the case for environmental protection advanced by environmental NGOs. It is therefore misleading to regard development as a purely material concern, and environmental protection as a moral concern, since development also forms part of a justice discourse for advocates from the South who see environmental protection as tied to interests of the North. This, at any rate, is how many countries in the South regard the North's increasing preoccupation with environmental concerns. As Henry Shue has pointed out, the goals of economic development and environmental protection are best understood as "interest grounded norms."[47] As Shue goes on to explain, "Serious ethics operates at the centre, not the fringe, of conceptions of legitimate interests."[48]

Developing countries have rightly argued that the developed world has achieved its relative affluence by exploiting the environment and fossil fuels (along with the environments and peoples of the colonies), yet it is now seeking, under the guise of the new environmental discourses, to "kick the ladder down behind it," that is, deny the South the same easy route to affluence without providing sufficient institutional capacity, resources, and technologies to make a genuinely alternative development path possible for the South. Against such a backdrop, the necessary precondition to any revision by the South of its negative sovereignty discourse—its categorical assertions of the right to self-determination and development—is not only greater distributive justice but also greater communicative justice, which would provide a greater parity of negotiating power between North and South in shaping the structure and direction of the global economy.

These arguments from the South have been partially recognized in the widespread endorsement of the general principle of "common but differentiated responsibilities" in the sustainable development debate in the lead-up and aftermath to the UN Conference on Environment and Development (UNCED) in 1992. This principle recognizes that not all countries are equally placed to pursue sustainable development, and that the developed world ought to take the lead. Nonetheless, unless the developed world provides greater capacity building assistance to developing countries, and provides more general measures to improve both distributive and communicative justice, then the developing countries are likely to continue to enlist a negative sovereignty discourse of nonintervention. Below I suggest how this negative sovereignty discourse might be used to promote, rather than resist, environmental protection.

8.2.4 Ecological Security and New Norms of Intervention?

The right to develop and the concomitant right of permanent sovereignty over natural resources were not the only norms to undergo significant reinterpretation in the post–World War II period. Since the end of the cold war the fundamental right of defense and the meaning of security, including understanding what constitutes a threat to security, how such threats ought to be contained, and what best fosters international order, have also undergone significant shifts.[49]

Traditionally the notion of security has been understood as a set of conditions that guarantees the ability of a state to pursue its national interests, free from both real and imagined impediments and threats. This has included the fundamental right of states to defend their territories from armed aggression. More recently attention has been directed to the idea that environmental problems (including natural resource shortages) may generate conflict between states.[50] Of course, many of these arguments pose no challenge to the traditional realist framing of state security.[51] However, the more ambitious lines of argument within the competing discourses of ecological security provide a fundamental challenge to the traditional referent of security (replacing the "state" with "citizen" and "ecosystem"), to the very idea of *territorial* defense, and to the means that might be deployed to secure the safety and integrity of citizens and ecosystems. Thus, instead of including environmental problems (e.g., global warming, deforestation, species extinction, and pollution) as a new range of threats to the national interest, to be met by military responses, such problems have been represented as posing a challenge to the state system, to the principle of territorial rule, to the activities and privileges of the military and the basic priorities of states, calling forth new forms of ecologically enlightened development, diplomacy, and governance.

Yet while everyone seems to recognize the political mileage to be gained from framing ecological problems in terms of security threats, there is a considerable unease within global political ecology circles about the wisdom of securitizing ecological problems in this way. That is, far from greening the state, the military, and global governance, it is feared that the concept of environmental or ecological security may serve to reinforce a Hobbesian state system and legitimate the militarization of state responses to environmental threats.[52] Thus Simon Dalby has pointed "to the contradictions of linking the politically conservative language of security with the radical action programs demanded by more far-thinking environmentalists."[53] As Daniel Deudney points out in his influential critique of the concept, military threats and environmental threats are of a different order and therefore should be dealt with by different agents and processes. Military threats are usually discrete, specific, and deliberate, involving a zero-sum game and an "us versus them" mentality. In contrast, environmental threats are usually diffuse, transboundary, unintended, operate over longer time scales, and implicate a wide range of actors ("ourselves" rather than "the other"), and their resolution usually carries common benefits. Moreover short-term, highly technical and military responses are rarely able to tackle the underlying causes of environmental problems.[54] Thus ensuring the sustainability of natural resources and the integrity of ecosystems is more likely to be achieved by negotiating fairer international trading and credit rules and better environmental treaties rather than enlisting military intervention.[55] Indeed, military responses would seem singularly inappropriate for dealing with the vast bulk of ecological problems. As Mark Imber has reminded us, "UNCED recognised the nearly irrelevant role of military force as a credible bargaining threat or sanction for non-compliance with environmental agreements."[56] Mathias Finger has taken this argument one step further in his claim that once the role of the military is critically confronted as a cause of rather than cure for ecological degradation, then disarmament emerges as one of the most urgent and significant steps toward global ecological integrity.[57]

Clearly, there are many dangers associated with a too hurried embrace of the concept of ecological security, and it would seem that critical political ecologists may be well advised to work within the already well-established discourse of sustainable development and ecological modernization. Yet Deudney's critique merely argues against linking environmental degradation with the concept of *national* security. He does not explore the potential of ecological security as a subversive discourse that might introduce new values and goals concerning, say, the inextricable links between the integrity of ecosystems and the sustainable livelihoods of people. This influential critique thus leaves unexamined the question as to whether there remains any legitimate *international* role for a green military operating, say, under UN auspices.

The expanding ideas of security, together with the idea that a community of democratic states provides the best guarantee of a peaceful international order, have become increasingly popular in recent times and have licensed an expanded role for the new generation of UN peacekeeping operations. In particular, the international community now appears more ready to support both humanitarian and military intervention to prevent gross human rights abuses. The shield of sovereignty can no longer be invoked by states to avoid condemnation and possible intervention in the areas of slavery, piracy, genocide, and colonialism. In all of these areas, one can point to ecological counterparts to these practices: the enslavement of animals in factory farms and scientific laboratories, biopiracy, the clearing of land containing threatened and endangered species, and the colonization of New World ecosystems with plants and animals from Europe. While I am not suggesting military intervention as a means of curbing these practices, they are nonetheless increasingly attracting both domestic and international censure. Moreover, in extreme cases, one can envisage circumstances when military intervention might be warranted to prevent what might be called crimes against nature or willful acts of ecocide (e.g., ecological terrorism and ecological sabotage). While there is no generally recognized right to environmental quality, there is plenty of scope within the existing repertoire of human rights to develop a case for humanitarian and, in some cases, military intervention to prevent, for example, direct threats to human health and physical security. It could be argued that there is now a need for a UN Ecological Security Council charged with the task of dealing with international ecological emergencies. Here one can envisage the deployment of UN troops (a veritable band of "green berets") to perform emergency services in defense of ecological security and those human rights that uphold the health and biophysical safety of citizens. The collective security system of the UN Charter confers on the Security Council the responsibility to "maintain peace and security" (article 24.1). Such proposals would appear to form the most likely basis for new forms of ecological intervention in the domestic affairs of states, either in response to deliberate crimes against nature or in response to major environmental disasters such as nuclear reactor meltdowns, major oil spills, and industrial chemical disasters (e.g., Bhopal).

Yet it might also be argued that such emergencies can be dealt with equally well by nonmilitary emergency services. Moreover there are very few ecological problems that constitute a genuine *crisis* in the sense of involving a high level of threat, a short period of warning, *and* the need for rapid military or para-military response.[58] It therefore is unlikely that ecological questions will ever play a big role in the deliberations of the UN Security Council and that, accordingly, the prospects of the development of any new ethics of ecological intervention are remote.

In this respect the critics such as Daniel Deudney and Jon Barnett are right: an ecologically sustainable world order is more likely to be achieved by the building of environmental capacity in the South, and by debt relief, poverty alleviation, and the development of more ecologically reflexive modernization. Thus it would be more fruitful for critical political ecologists to focus on what might be more general norms of ecologically responsible statehood than seek to securitize ecological problems or develop norms of military intervention.

Although not all ecological problems are amenable to securitization, there is always an ecological dimension to the role of military. The use of both conventional weapons and weapons of mass destruction—not to mention military training and weapon production, storage, and disposal—have all contributed to environmental degradation, as was noted in chapter 2. Some of this damage is incidental or so-called collateral damage. However, much of it is also intentional, such as the willful destruction of ecological assets and natural resources by nation-states (e.g., the US military's use of Agent Orange to destroy forest cover and food crops during the Vietnam war). Against the background of the 1990–1991 Gulf war, particularly the massive burning oil wells in Kuwait, the Earth Summit negotiations included a discussion of the idea of a "green protocol" to the Geneva Red Cross Conventions.[59] Although this argument did not succeed, the negotiating states agreed in the Rio Declaration to acknowledge that "warfare is inherently destructive of sustainable development" and to call upon states to "respect international law providing protection for the environment in times of armed conflict" (principle 24 of the Rio Declaration).

Against this general background, it is now possible to explore how the new normative discourses concerning the rights and responsibilities of states might be coaxed in an even greener direction. Although I have suggested that there is little mileage in developing an ethics of intervention in the name of environmental protection, I will attempt to show that there may be considerably more mileage in developing an ethics of nonintervention to uphold the ecological integrity of state territory and enforce the practice of ecologically responsible statehood.

8.3 Ecological Harm, Nonintervention, and Ecologically Responsible Statehood

In the introduction I argued that evolutionary change to the *third* face of sovereignty (i.e., the rights and responsibilities of states) was where the most significant ecological developments have taken place. In terms of the four levels of analysis singled out, it has been the expansion in the number, scope, and sophistication of environmental regimes, declarations, and strategies where the biggest changes have occurred. However, I have suggested that these developments are best understood as evolutions rather than revolutions in sovereignty. More generally, the institution of state sovereignty has endured against a background of an evolving set of constitutive discourses concerning self-determination, development, human rights, and security, all of which have undergone some modest greening over the past thirty to forty years.

Now it has been argued, by Steven Bernstein, that the shifts in international environmental law and policy over the last forty years represent an uneasy compromise between the increasingly influential neoliberal economic discourse, on the one hand, and environmental protection, on the other hand. What Bernstein has called "the compromise of liberal environmentalism" in international environmental regimes is one whereby environmental arguments have been mostly adapted to, and absorbed by, preexisting neoliberal ideas concerning the virtues of competition and unfettered economic exchange.[60] According to Bernstein, this liberal compromise—which is reflected in the Rio Declaration, agenda 21, and the UN Framework Convention on Climate Change—has enabled environmental ideas to become more prominent in international legal and policy circles than they otherwise might have been. The same might be said for environmental ideas and the institution of state sovereignty.[61]

Turning to the domestic level, one might be tempted to concur with Bernstein given the weak (i.e., mostly technocratic) ecological modernization that has occurred in most OECD states. As I argued in chapters 3 and 4, weak ecological modernization reinforces rather than challenges the processes of modernization, technological change, and the instrumental exploitation and management of nature.

Yet the compromise of liberal environmentalism at both the international and national levels contains continuing and significant tensions. On the one hand, the compromise is predicated on continued economic growth within the context of states, capitalist economies, and the increasing use of market-based instruments for environmental protection (which represent an application of neoclassical economic prescriptions for the "internalization" of negative ecological externalities and the more efficient allocation of resources).[62]

On the other hand, the international acceptance of principles such as the precautionary principle and the principle of common but differentiated responsibilities gives expression to stronger moral norms of environmental protection and environmental justice that have a much less obvious connection to the Washington (neoliberal) consensus. Similarly the shift from indiscriminate growth to weak ecological modernization in many OECD countries is part of a process of critical questioning and adaptive policy learning that has the potential to bring about a challenge to not only the means but also the societal goals and the purposes and character of the social steering systems designed to deliver those goals. After all, environmental ideas and the environmental movement have developed *after* the establishment of capitalism and the sovereign state. It is hardly surprising, then, that those ideas that appear to fit the normative framework of preexisting institutions are more likely to be successful than those that challenge it, as Bernstein argues. Nonetheless, the fact that some of these new norms have begun to take hold despite the fact that they do not easily nest within the preexisting discourses of development and the institution of sovereignty provides an opening to reshape these discourses and institutions.[63] The questions to which I turn are: How might these tensions be exploited, what opportunities for further greening are imminent in these developments, and how might they be framed and pursued?

What is most interesting for our purposes is the ways in which *both* positive and negative discourses of sovereignty have been enlisted to work both for and against the environment. These developments are not dissimilar to the way in which the concept of autonomy has been fleshed out by political theorists in positive and negative terms.

Take, for example, the principle of permanent sovereignty over natural resources and the broader principle of the right to develop. Although this principle asserts a positive right on the part of sovereign states to promote the economic welfare of their citizens, the aspect that is emphasized in the developing countries' construction of this international discourse is the assertion of freedom from outside interference by NGOs, multilateral lending agencies, and Northern states. The principle of permanent sovereignty over natural resources has thus been used as a shield against unwelcome environmental encroachments into the newly acquired powers of statehood. Such encroachments are seen merely as a continuation of the legacy of imperialism under a green guise. Thus new international environmental regulations, such as those associated with combating climate change, are seen as reaching directly into the basic policy-making prerogatives of states (e.g., energy and fiscal policy) while putting developing states at a further disadvantage relative to developed states (who were able to achieve their economic dominance by means of indiscriminate growth in the past). Needless to say, whereas the developing countries' resistance to such regulation is, for the most part, a *principled* resistance, many developed countries—notably the United States and Australia—have resisted international environmental cooperation on the ground of more thinly veiled economic expediency. Against this background of intransigence on the part of the South and the North, it is hardly surprising that many international environmentalists and global political ecologists continue to regard the institution of state sovereignty as a major obstacle to enlightened global environmental governance.

However, it needs to be asked why this negative sovereignty discourse has not been more frequently invoked to argue that unwanted ecological risks and problems that flow *into* the territory of states from outside represent an unwarranted interference or encroachment in the territories and affairs of nation-states by the relevant culprit states. Such a move is by no means foreclosed by my defense of the transnational-state in the previous chapter, which retains and builds upon (rather than rejects) the principle of belongingness/membership in a political community and welcomes the development of local environmental patriotism at the same time as it seeks to cultivate a cosmopolitan practice of ecological citizenship. One of the purposes of that chapter was to explore how the democratic legal system might be made more effective as a social steering system vis-à-vis other steering systems, notably the economy. Here I seek to show how the two sides of the sovereignty coin—self-determination and nonintervention—can be made to work for democracy and the protection of local ecosystems.

For example, there is good ground for arguing that atmospheric, marine and aquatic pollution, hazardous waste, and genetically modified organisms entering the territory of a particular states as a result of activities sanctioned by other states constitute a form of illegitimate intervention in the ecosystems of states. In the context of increasing global economic and ecological interdependence, such unwanted flows of pollution (e.g., acid rain), waste, or potentially harmful products might also be said to undermine the self-determination of nation-states, in this case, the freedom of national communities to determine their own levels of environmental quality and the ways in which they might wish to use sustainably or protect their natural resources and biological and cultural heritage.

More generally, certain forms of transboundary pollution or trade in wastes or unsafe goods might in certain circumstances also constitute a health and security threat to citizens such as to amount to an infringement of human rights (e.g., radiation, contaminated water, and food supply). Principle 1 of the Stockholm Declaration recognized the links between environmental protection and human rights and the World Commission on Environment and Development strongly endorsed the idea that "All human beings have the fundamental right to an environment adequate for their health and well-being."[64] More recently the Draft Declaration of Principles on Human Rights and the Environment commissioned by the United Nations Commission on Human Rights could form the basis for the negotiation of a more formal, binding international instrument.[65]

Such an argument would build on, and extend, the principle of nonintervention embodied in the United Nations Charter, which provides that "All members shall refrain in their international relations from the threat or use of force against the *territorial integrity or political independence* of any state, or in any other manner inconsistent with the Purposes of the United Nations" (article 2.4). Here territorial integrity could be interpreted to include "ecosystem integrity" and linked to the emerging discourses of ecological security and environmental rights. Article 1 of the Charter lists the primary purposes of the United Nations as maintaining international peace and security and upholding human rights.

In view of the untapped ecological potential of this negative sovereignty discourse, it is necessary to question the popular environmental argument that exclusive territorial rule is fundamentally incompatible with sound ecological management. Indeed, territoriality is of secondary importance to the other dimensions of sovereignty when it comes to exploring the potential for *further* ecological revisions to sovereignty. This is because sovereign territorial rule is not necessarily ecologically problematic *if it is contextualized and qualified* by, say, ecological standards of membership, ecological standards of democratic legitimacy, or new ecological rights and responsibilities of states. Indeed, I have suggested how the territorial dimension of sovereignty might be made to work *for* the environment insofar as it might enable political communities to resist certain aspects of economic globalization or to censure and possibly also halt practices in other states that threaten to undermine domestic ecosystem integrity. In these circumstances at least territorial sovereignty—including self-determination and the associated principle of nonintervention—can serve as a bulwark *against* anti-ecological practices that encroach upon the territory and policy-making powers of particular nation-states.[66]

The negative sovereignty discourse is already implicit in many environmental treaties, including two treaties that have been actively promoted by developing countries: the Basel Convention on the Transboundary Movement of Hazardous Waste and the Cartagena Biosafety Protocol negotiated under the Convention on Biological Diversity. In both cases, developing countries have been in the forefront of negotiating legal norms that are intended to protect their environment and biodiversity from hazardous waste or genetically modified organisms produced mainly by transnational corporations based in the developed world. (Some subnational ethnic groups have also added an environmental twist to their claims for self-determination. Indigenous peoples, in particular, have mobilized the *human* right to self-determination to defend their attachment to local environments and to resist the development priorities of settler states under postcolonialism.[67])

Despite the frequent claim that environmental multilateralism has become a form of green imperialism inflicted on the South by the North, these examples demonstrate how the negative sovereignty discourse—cashed out in terms of environmental treaties that seek to restrict trade in environmentally harmful wastes or products—can be used as a green shield by the South. There are doubtless considerably more ways in which the discourse of nonintervention might be developed as a form of ecological resistance, particularly in the face of the relentless processes of trade liberalization or in the face of nuclear or chemical weapon installations in neighboring states.

However, the principle of state responsibility for environmental harm has not developed to the point where unwanted incursions of ecologically harmful substances into the territory of states have earned the status of an "unjustified intervention" or territorial transgression, as a matter of general principle. Nonetheless, the argument has been tried in the courts. For example, in its submission against France's atmospheric nuclear testing in the Pacific, Australia argued that France's conduct represented an "intrusion upon Australian territory of foreign matter" (i.e., ionizing radiation), which constituted an infringement of Australia's sovereignty, in terms of the integrity of its territory and the "decisional aspect of its sovereignty" (i.e., its sovereign right to make its own decisions about what should happen in its territory).[68]

However, there seem to be two likely objections against developing the negative sovereignty discourse along these lines. First, it might be said that a negative ecological sovereignty discourse is unlikely to gain much support from states, since it requires a major shift in emphasis in the environment/development debate that runs contrary to the basic development interests of states. Second, it might be pointed out that a new negative ecological sovereignty discourse merely serves to reinforce the institution of sovereignty, when the point should be to challenge and undermine it. In other words, the focus on re-envisioning, or ecologizing, sovereignty is misplaced, and critical political ecologists should be working to develop alternatives to the principle of exclusive territorial rule.

Turning to the first argument, the flip side to the assertion of any particular right of environmental noninterference carries with it a correlative set of responsibilities on the part of other states to refrain from allowing such encroachments. In effect, such a discourse would seek to impose on states a duty not to carry out any activities that would cause environmental harm to any other state, since such harm would be considered a wrongful encroachment on sovereignty. Since most economic activities produce some negative ecological externalities, many of which are also transboundary and not all can be foreseen, fulfillment of such a duty would require some drastic changes to patterns of investment, production, and consumption in most states in order to prevent and minimize such externalities. In effect, such a duty would radically transform and enlarge the roles and identities of states to include the roles of ecological protector, trustee, and risk minimizer. Moreover, if the precautionary principle were to be put to work as an evidentiary rule in cases of scientific uncertainty, then the rights of environmental victims would take precedence over the rights of territorial use, including the rights of public and private property holders. That is, once a prima facie case is raised by victim states that serious or irreversible environmental harm has occurred or might reasonably occur, then the onus would shift to the defendant state to show that the offending economic activities within its jurisdiction and control have not or would not cause such harm. Proving the absence of likely harm is always a more difficult evidentiary burden than proving the existence of harm.

In support of this reconstruction, it might be argued that the broader trend in international environmental law and policy is already pointing away from a construction of the state as owner/overlord of its territory and toward that of caretaker/trustee of territory, with custodial and management obligations owed not only to other states but also to citizens and the global community. Once this shift is recognized and consolidated, the argument for the systematic application of the precautionary principle in relation to transboundary ecological risks appears less provocative. That is, if states are understood to have a positive obligation to act as environmental caretaker as part of their very raison d'être, then a case can be made for the replacement of the traditional presumption in favor of territorial use (or environmental exploitation, subject to qualifications) with a presumption in favor of environmental protection and the prevention of victimisation, at least in those circumstances where serious or irreversible environmental risks are at stake.

Of course, the development of a green sovereignty discourse of this kind that imposes direct and enforceable environmental obligations on states to refrain from certain activities that *might cause serious harm* to the environment is likely to be vigorously resisted by most states, not to mention the corporate world. While the right to develop has been qualified in the Stockholm and Rio declarations by the responsibility of states "to ensure that activities within their jurisdiction or control do not cause damage to the environment of other States or of areas beyond the limits of national jurisdiction," these new responsibilities merely qualify the presumptive right in favor of territorial use. That is, the right to develop (including the right of permanent sovereignty over natural resources) is still considered to be a central component of the right of self-determination and a primary end of the state, but the *means* by which this end is to be pursued is increasingly limited by a growing array of environmental responsibilities. On this construction it would seem that the fulfillment of such environmental responsibilities still remains incidental to the primary purpose of the state (security and the pursuit of economic development); such environmental responsibilities do not themselves constitute part of the raison d'être of the state. They are mere side constraints. Against this, my reconstruction seeks to recast the state's environmental responsibilities by recasting the fundamental purpose of states.

Not surprisingly, the few attempts in the international ethics and political theory literature to develop new norms of ecologically responsible statehood have tended to follow the conventional grain of analysis or else seek alternatives to sovereignty rather than fundamentally challenge the presumption in favor of territorial use.[69] For example, Henry Shue has sought to articulate stricter ecological qualifications to the sovereign right of states to develop in terms of the well-established liberal harm principle, which does not seek to overturn the presumption in favor or territorial use. Using the analogy of just war theory, Shue has argued that the state's economic development prerogatives are a "just cause," but they must not be promoted by "unjust means." Accordingly, there ought to be *binding external limits* on the means by which domestic economic goals are pursued. That is, it is just as wrong for a state to slaughter innocent people in a just war as it is for a state to inflict serious harm on innocent foreigners in pursuing its otherwise laudable goal of economic development. (In defending the idea of side constraints this argument is structurally similar to Marcel Wissenburg's arguments for qualifying property rights, as discussed in chapter 4).

The fundamental point is that while, according to Shue, it is permissible for states to advance the interests of their own nationals ahead of foreigners, it is not permissible to do so in ways that cause serious harm to foreigners where such harm can be reasonably avoided by alternative development policies. Clearly, these environmental duties do not seek to dislodge the centrality of the state's development prerogatives or control of territory; they merely seek to qualify the means by which states might pursue their otherwise legitimate right to develop.[70] Nonetheless, Shue is quite explicit in insisting that states take environmental responsibility for everyone affected (rather than just citizens) in certain circumstances:

- If their policies contribute substantially to the harm suffered by outsiders.
- If the states in which the victims live are relatively powerless to stop the harm.
- If the harm is to a vital human interest like physical integrity.
- If an alternative policy is available that would not harm any vital interest of anyone inside or outside the state.[71]

Shue's proposals would clearly extend the norms for state responsibility for environmental harm beyond the existing customary law and provide something approaching the ideal of ecological democracy defended in chapter 5. However, it still assumes that victim states will have the onus of proving actual harm, causality and lack of due diligence (or in this case, failing to implement alternative, nondamaging policies) on the part of defendant states. The question of the prevention of harm (as distinct from compensation for harm) is not explicitly addressed, so it is still not clear whether these principles might have enabled Australia to seek an injunction against, or otherwise seek to preempt, the French nuclear tests rather than wait for the damage and seek compensation. In this respect Shue's proposals are closer to the polluter pays principle than the precautionary principle.

In view of the likely resistance to the more radical negative environmental sovereignty proposal outlined above, Shue's case for working with well-entrenched liberal principles, such as the harm principle, in the context of existing presumptions and precedents is likely to be the more successful strategy in the short term. Yet it might also prove to be quietly subversive in the medium and longer term as the range of ecologically sustainable development strategies (pursued by greener states) proliferate. Under such circumstances defendant states with adequate environmental capacity will find it increasingly difficult to show that there are no feasible alternatives to the more conventional economic practices that generate the environmental harm under dispute.

To the extent to which the processes of ecological modernization and environmental democracy deepen along the lines envisaged in chapters

3 and 5, then one might also see deeper shifts in shared expectations and understandings about the role and purpose of states such that protecting the environment is recognized as a no less fundamental role of states than promoting development. Among green democratic states, at least, this would not represent merely a functional adaptation of sovereignty to economic and ecological interdependence. Rather, it would also represent constitutive change entailing shifts in the self-understanding and roles of property owners and sovereign states alike. That is, to the extent to which green democratic states have moved beyond the "unthinking dogmas of liberalism" outlined in chapter 4, then the environment would no longer be understood in purely instrumental terms and the practice of sovereign rule would no longer be understood to be so crucially dependent on the rational (meaning more efficient) mastery of nature. Instead, states would be understood as internally related, and the notions of self-rule would be understood as constituted (rather than restricted by) the web of shared understandings, relationships and multilateral norms that give meaning to, and make possible, the authority and practice of government.[72]

Of course, the sorts of constitutive changes that I have sketched are a very long way off and the first objection would therefore seem to hold to the effect that states would be far too resistant to such changes. As Peter Penz has perceptively pointed out, the current failure of most states to prevent environmental victimization *within* their borders will also weaken any commitment to prevent environmental harm *beyond* their borders.[73] As Penz also argues, the reason the right to noninterference has not been given an environmental interpretation is because *using* the environment for production is still more central to the interests of states than guarding or protecting it.[74]

The argument that environmental protection and preservation run counter to the main security and economic imperatives constituting states is one that is common to both neo-realist and neo-Marxist analyses (as explored in chapters 2 and 3). Many global political ecologists tend to share this analysis. According to Ken Conca, if one takes the deep structure of global politics to refer to economic globalization, sovereignty and the associated discourse of modernity, then there is little evidence of any deep structural transformation brought about by international environmental politics.[75] In short, as Conca puts it, "*Our Common Future* is a text on sustainable development—but sovereignty, modernity and capitalism form its subtext."[76] Accordingly, for those who regard states as irredeemably tied to productivist interests, exploring alternatives to the institution of sovereignty becomes more compelling. For his part, Penz ultimately recommends environmental federalism, based on the principle of subsidiarity, over the existing confederal state system, as a superior governance structure for ecological management. Included in this new arrangement would be a global environmental protection authority (democratically represented by citizens of the world, not states) charged with the task of preventing international environmental harm.[77] However, given that environmental change is mostly shaped by economic rather than environmental policies, then such an authority would need to be empowered to reach into the decision-making processes of the various institutions of economic multilateralism (the WTO, the World Bank, and the IMF). Similarly Nicholas Low and Brendan Gleeson recommend as the first steps toward a "world constitution for environmental and ecological justice" the establishment, under the mantle of the United Nations, of a World Environment Council, and a World Environment Court. The World Environment Council would represent citizens, not states, and be charged with the task of discursively developing global principles of environmental justice and law that would presumably strengthen the existing "negotiated order" of multilateral regimes.[78]

Yet the problem with these proposed alternatives to the state system is that they presuppose agreement by states. Moreover, in the absence of any such agreement, such calls begin to look like "a counsel of despair," to borrow again Hedley Bull's phrase.[79] This chapter has drawn attention to the series of not inconsiderable "evolutions in sovereignty" in terms of the environmental rights and responsibilities of states, and the prospects of further such evolutions appear a good deal brighter than the likelihood of any decline of sovereign states or any movement beyond the state system.

In any event, working on the more urgent and finer grained tasks of improving domestic environmental policy in specific subject domains and promoting stronger ecological modernization generally, including finding ways of both reducing and internalizing ecological costs produced by firms and public agencies, is nonetheless to *point* in the direction intimated in this chapter. As I have suggested in the previous chapter, the accumulation of such bottom-up, unit-level transformations provide the most likely way in which more general, system level transformations will take place.

Conclusion: Sovereignty and Democracy Working Together

The anarchic state system, global capitalism, and the administrative state have served in different ways to inhibit the development of greener states and societies. In this book I have shown how three mutually informing counterdevelopments—environmental multilateralism, ecological modernization, and the emergence of green discursive designs—have emerged to moderate, restrain and in some cases transform the anti-ecological dynamics of these deeply embedded structures. The counterdevelopments have been brought about by both state and nonstate social agents acting to enhance the state's receptivity to ecological concerns, and its capacity for social and ecological learning.[1] In short, the mutual democratization of states and their societies appears to operate in a virtuous relationship with more reflexive ecological modernization at the domestic level and more active environmental citizenship by such states (and their citizens) on the international stage, when compared to other states.

This virtuous relationship, however, cannot be deepened without a move from liberal democracy to ecological democracy. Despite increasing environmental regulation, liberal democratic states have been unable to resolve or significantly minimize many ecological problems, and they continue to permit the displacement of ecological costs over space and time. To suggest that this state of affairs is merely a reflection of peoples' preferences or that it is the price of upholding human autonomy and tolerating moral pluralism on environmental matters is far too complacent a view.

Liberal democracies continue to construct decisions to invest, produce, and consume as essentially private matters, unless such decisions can be shown to cause direct and demonstrable harm to identifiable agents (which is never an easy matter). This construction serves to depoliticize decision making in the very domains that generate diffuse yet cumulative ecological impacts. The liberal focus on the *individual's* freedom to choose in the sphere of politics, economics, and lifestyle (along with the corporation's freedom to invest) deflects attention from the social and economic structures that shape and limit the horizons of individual choice (including environmental choices), and more so for economically marginal social classes and groups. When the liberal democratic state permits social actors to displace ecological costs on to others, it restricts the ability of environmental victims (both inside and outside the borders of the nation-state) to enjoy the full range of freedoms that liberalism supposedly upholds, including the freedom to participate or otherwise be represented in the making of decisions that bear upon their own lives.

In chapter 4, I argued that the inability of the liberal democratic state to provide systematic environmental protection can be traced to the bourgeois origins of liberalism's conception of autonomy and to a range of associated "liberal dogmas" that would not survive the critical scrutiny of a genuinely unconstrained and inclusive communication community in the contemporary, deeply interconnected world. Liberalism's atomistic ontology of the self, its quest for mastery of the external world through the application of instrumental reason, and its corresponding denial of any noninstrumental dependency on the social and biological world have ultimately imperiled rather than enhanced human autonomy for many and environmental integrity for all. By sheltering these articles of faith from further critical questioning, liberalism has lost sight of the dependence of autonomy on critique and thwarted the realization of autonomy for a much wider constituency than is currently the case.

In chapter 5, I defended an ecologically renovated, postliberal democracy that could confront such liberal dogmas, and call social agents to account for their risk generating decisions. Ecological democracy would differ from liberal democracy in enabling more concerted political questioning of traditional boundaries between what is public and private, domestic and international, intrinsically valuable and instrumentally valuable. It would be concerned to maintain the ongoing contestability of public and private power by means of inclusive representation and critical deliberation in relation to risk generating decisions. The regulative ideal or ambit claim of ecological democracy is that all those potentially affected by ecological risks ought to have some meaningful opportunity to participate, or be represented, in the determination of policies or decisions that may generate risks. I have sought to illustrate how this ambitious ideal can nonetheless be practically embodied in the constitutional framework and due processes of the green democratic state. Drawing together these arguments it is possible to offer a broad sketch of what might be included in a green constitution.

For example, at the broad level of constitutional purpose, the green democratic state would be outward looking rather than parochial or nationalistic, reflected in a preamble that includes a commitment not only to human rights but also a statement of responsibility to protect biodiversity and the life-support services and integrity of the earth's ecosystems.

In terms of substantive provisions one might envisage a charter of citizens' environmental rights and responsibilities as part of the standard list of civil and political rights (e.g., the right to vote, to stand for office, for fair and democratic elections, freedom of speech, assembly, association; freedom from discrimination; and inclusive political representation and participation). This additional cluster of substantive and procedural environmental rights and responsibilities, might include the following:

- A right to environmental information (backed up by mandatory state of the environment reporting, and community right-to-know legislation in relation to pollutants and other toxic substances)
- A right to be informed of risk-generating proposals
- A right to participate in the environmental impact assessment of new development and technology proposals
- A right to participate in the negotiation of environmental standards
- A right to remedies when environmental harm is suffered or threatened
- Third-party litigation rights to enable NGOs and concerned citizens to ensure that public environmental laws, including minimum environmental standards, are being upheld
- A responsibility on the part of all state decision makers and corporations to adopt a cautious approach to risk assessment (this might be

expressed in terms of the constitutional entrenchment of the precautionary principle, which is enlarged to include future generations and nonhuman species as moral referents)

• A responsibility to avoid, or where necessary pay compensation for, causing any environmental harm to innocent third parties (this might be expressed in terms of the constitutional entrenchment of the polluter-pays principle)

• The constitutional entrenchment of an independent public authority—such as an environmental defenders office—charged with the responsibility of politically and legally representing public environmental interests, including the interests of nonhuman species and future generations

• The provision of constitutional authority for the holding of cross-border referenda and reciprocal representation in deliberative forums in relation to matters of transboundary or common environmental concern with citizens of other states in those circumstances where they may be seriously affected by proposed developments taking place within the state (the activation of this authority would require reciprocal agreements with other states)

• In the case of federal states, clear and unequivocal legislative powers to protect the environment by the national or central government to prevent evasion of both domestic environmental responsibility (i.e., by means of buck passing to provincial units) and to facilitate the swift enactment and implementation of environmental treaties

These are merely suggestions illustrative of the possibilities for greening existing democratic constitutions. Clearly, the green democratic state is not a neutral state, but then again, nor is the liberal democratic state. Both shape and reflect different conceptions of moral and political community.

The constitutional renovations I have sketched build upon, rather than reject, the liberal and republican legacies of constitutional democracy, the rule of law, the separation of powers, and the accountability of the executive to parliament and the public. However, the additional range of substantive and procedural environmental rights and decision rules would secure more systematic consideration of a much wider environmental constituency than just the citizens of the nation-state. Moving toward this ideal requires that states become "local agents of the common good" (to adopt Hedley Bull's phrase),[2] facilitators of transboundary democracy and therefore increasingly "transnational" in their orientation. The purpose of the green constitution would be to provide a structure of government that enables, and where necessary enforces, ecological responsibility on behalf of the broader community at risk. It provides the fundamental reframing of the purpose of the state, together with the democratic principles and procedural requirements that confer legitimacy on the state.

However, the ultimate success of the green democratic state should be measured not simply by the appearance of constitutional renovations and democratic procedures of the kind that I have suggested. Rather it should be measured by the changes in the economy and society that it has helped to facilitate or, as Lennart Lundqvist has put it, when "*ecological* evaluations are so internalised and integrated that they become as 'natural' as more conventional *economic* terms actors presently apply whenever they make decisions as producers or consumers."[3] The green public sphere is absolutely crucial in facilitating this broad cultural shift toward an ecological sensibility, in the same way that the bourgeois public sphere facilitated the shift toward the widespread diffusion of liberal market values.

Constitutional change, then, is merely a necessary, not a sufficient condition, for the greening of the economy and society. Although the primary focus of this book is on the state, my arguments also presuppose active ecological citizens that take responsibility for their state as *their* creation, and bring to life the kinds of green constitutional reforms that I have recommended. This requires not simply resisting the state when it sanctions ecological destruction or engages in various forms of social domination, but actively reshaping and monitoring it to ensure that it promotes ecological sustainability and social justice. Moreover constitutional design is only one facet of the environmental capacity building that is required for states and their societies to respond to ecological problems in a more concerted fashion. Environmental capacity building encompasses the structural preconditions for state-societal solutions to ecological problems. This includes knowledge (ecological, technological, vernacular, and administrative), legal and material resources, effective policy-making institutions, innovative industry, and political participation, underpinned by an active civil society and critical public spheres, including green public spheres.[4] Indeed, the greening of the constitution along the lines suggested might well be the *culmination* of a series of normative and material shifts toward sustainability emanating from elsewhere—within civil society, the economy, and in domestic and international public policy—rather than the primary *cause* of such shifts. Ultimately, however, the green constitution, the green economy, and the green civil society (and public sphere) would become mutually reinforcing. For example, the inclusion of a charter of rights and responsibilities in the green constitution could provide a catalyst for moving from simple modernization to more reflexive modernization, from the mere pursuit of improvements in environmental productivity to the more concerted pursuit of environmental justice.

Such a virtuous circle of change would not take place without critical green public spheres, which are essential if the state and the economy are to be constrained by the needs of the lifeworld. It is primarily through the process of critical questioning and reflection in public spheres that it becomes possible to rethink not simply the means by which society pursues established goals but also the goals as well. At the same time the political transition toward a green democratic state requires not just the proliferation of critical green public spheres but also politically oriented green *movements*, that is, the political marshaling of arguments, and the mobilization of people, organizations and political parties, for change. The green movement, considered as a whole, is a broad, decentered, and heterogeneous movement made up of new social movements (e.g., environment, peace, anti-nuclear, aid, poverty, Third World development, and women's), new political parties, scientists, research institutes, environmental educators, journalists, ordinary citizens, and ecologically modernizing firms. Moreover this movement has produced a network of crisscrossing and overlapping green public spheres that reflect a continuing debate about both the means and ends of ecologically sustainable development. While green parties remain the most obvious transmission belt for channeling this debate from civil society to the state, they are by no means the only one. In any event, green parties are unlikely to survive without the complementary, and sometimes conflictual, interplay of green political actors and their diverse networks of political communication.

Of course, the green movement faces considerable resistance from a variety of social actors and organizations whose material interests and/or political ideals are threatened by the case for a more ecologically sustainable society and economy. However, as with green arguments for change, it is always necessary to subject this resistance to critical scrutiny, to test its legitimacy in unconstrained public forums. Given the ways in which existing unequal power relations tend to thwart fulsome policy debate and risk evaluation, it is necessary for the green movement to develop strategies of empowerment for systematically excluded groups to achieve more inclusive societal deliberation. The case for deeper ecological reform is thus dependent on extending representation and deepening democratic participation to enable critical publicity and deliberation.

To the extent that serious ecological problems persist (and the distribution of actual environmental harm and potential risks remains highly skewed), then public anxiety will remain high, public trust in experts and managers will wane and political agitation or disaffection within civil society will grow. Such a situation creates the conditions for the marshaling of arguments for change. The intractability of ecological problems also provides the opening for arguments that go beyond mere technical fixes toward new policy principles, paradigms, and societal goals, and new understandings of the role and rationale of the state.

A central concern of this book has been to loosen, if not totally dislodge, the tight nexus between citizenship, democracy, territoriality, and sovereignty that is central to the regulative ideals of the liberal democratic state. Transboundary environmental problems provide a graphic illustration of the ways in which the principle of exclusive territorial rule works to restrict the boundaries of the moral community and thwart the further development of sovereignty, democracy, and citizenship. But territorial rule need not be exclusive in this way. In seeking to extend the boundaries of the moral and political community and transcend the traditional nexus of citizenship, democracy, territoriality, and sovereignty, chapter 7 has sought to work creatively with the enduring tensions between communitarian and cosmopolitan principles for ordering democratic life. In terms of regulative ideals, the green democratic state has been defended as a transnational state that enjoys the confidence of its own citizens and, on certain occasions involving reciprocal agreements, that of other communities that it may serve or assume responsibility toward in situations of common or transboundary ecological concern. Rather than defend an abstract, global, cosmopolitan democracy of the kind envisaged by David Held, I have suggested that it is more desirable and feasible to transnationalize democracy in piecemeal, experimental, consensual, and domain-relative ways. As Jürgen Habermas has argued, the enactment and effective implementation of legal norms requires finite space-time coordinates of the kind currently provided by the legal system of the territorial state. However, I argue against Habermas in suggesting that there is no reason why domestic legal systems cannot be enlisted to serve transboundary communities when such communities—either directly or through their representatives—are given the opportunity to be both the authors and addressees of common transboundary norms. Under such circumstances the principle of sovereignty can be brought into line with the requirements of democratic legitimacy and therefore operate in more inclusive ways. This is the sense in which ecological democracy can be made to work with, rather than against, state sovereignty at the institutional level.

Thinking of the green democratic state as a transnational state rather than just a nation-state does not obliterate the importance of national communities (which are increasingly multicultural in character). Rather, my argument is one that acknowledges and self-consciously seeks to build upon, rather than reject, the social bonds of existing national communities (and subcommunities) and the communitarian arguments about the importance of belonging and membership in particular communities and attachment to particular places. This strategy of working from the particular to the general also leaves room for retreat to the particular to enable the sovereignty discourse to be enlisted in environmentally defensive ways. That is, if changing patterns of production, trade and financial movements are leading to flows of unwanted and unwarranted incursions of pollution or environmentally unsafe products into particular communities, then such flows ought to be resisted on the grounds that they undermine the territorial integrity and right of self-determination of national communities to determine their own social conditions, levels of environmental quality and natural heritage protection. In some circumstances such flows (i.e., radiation) can take on the character of health and security threats.

Of course, the flip side to this green enlistment of sovereignty as a shield is a set of responsibilities on the part of all states not to cause environmental harm to other states. The problem is that the existing customary law principle of state responsibility for environmental harm provides only a weak protection against environmental victimization and little deterrence against the displacement of ecological costs by state and nonstate actors. Just like the common law presumption in favour of property holders, the presumption in favour of territorial use/development leaves it to victims to prove damage, causation and lack of due diligence in the context of prevailing standards of what is reasonable. Such presumptions merely qualify, but do not transform, the traditional role of states in promoting development. Nonetheless, I have sought to emphasize that the trajectory of international environmental law and policy is now at least pointing toward the idea of states taking on the new role of ecological stewards or trustees, with responsibilities toward not only their own citizens but other states and the global commons. Whether this trajectory is followed to the point of replacing these traditional presumptions with something like the precautionary principle is, of course, an open question.

Such a trajectory is beyond the radar screens of most mainstream international relations theorists. Implicit in the neorealist argument that states interact in a Hobbesian universe is the expectation that states will engage in the unrestrained exploitation of natural resources, species and ecosystems both within their own territory and also, where possible, beyond (e.g., by supporting bioprospecting in biodiversity rich, postcolonial territories). Regardless of whether neorealists would personally endorse environmental exploitation, their analysis nonetheless constructs such a posture on the part of all states as natural and inevitable. Moreover neorealists would expect that states would have no interest in securing the environmental protection of areas lying outside their own territories, or the territories they may control—other than for geopolitical purposes. Any qualifications to these dynamics can only be expected where they might contribute to the making or maintaining of strategic alliances with other states.

By contrast, neoliberal institutionalists, from their Lockean universe (where states relate to each other as respectful but calculating business rivals rather than hostile enemies), promote environmental regimes and domestic policies that instigate the wise use or more rational exploitation of natural resources such as arable land, water, species, timber, energy, and minerals. At their best such policy regimes would protect ecosystem services such as the waste assimilation capacity of ecosystems to maintain natural sources/resources and sinks for the production process. Already we have seen the increasing promotion of market-based measures for environmental protection, since these are reputed by economists to allow states to achieve desired environmental outcomes in a more flexible and cost-efficient manner. However, neoliberals would not expect states to pursue the protection of ecosystem services or the preservation of ecosystems or species for their own sake, or merely for their aesthetic or cultural value, unless these arguments can be disingenuously rendered in utilitarian terms from a self-interested state perspective (e.g., for the growing eco-tourist market). These utilitarian arrangements are certainly more promising than the resource exploitation expected by neorealists, and they have become increasingly influential in environmental policy circles. Yet from the critical political ecology perspective defended in this inquiry, they are only halfway along the way toward reflecting the full range of ecological and cultural arguments for the protection and preservation, as distinct from sustainable utilization, of natural resources, species, and ecosystems. Regardless of whether neoliberals may personally desire a greener world, their analysis assumes that states will not seek to move toward such a world unless it can be shown to be in their interests (understood in mostly economic terms). Thus a purely utilitarian posture towards the nonhuman world is naturalized by neoliberal institutionalists.

Against these mainstream approaches, I have argued that to the extent to which states can be found to operate in a Kantian or post-Westphalian culture of anarchy at the regional or international level, the possibilities for the creation of more innovative multilateral regimes for environmental protection are greatly enhanced. We have seen this in the European Union, most graphically with the Aarhus Convention, which effectively creates transboundary environmental justice rights. I have argued that the more the multilateral communicative context becomes both deliberative and inclusive the more likely it will be that environmental decisions will not only be environmentally prudent but also environmentally just. This is not an entirely fanciful wish but rather an emerging potential that is grounded in contemporary historical developments, aided by the fact that environmental NGOs, scientists, and the media are playing an increasingly significant role in the development and critical monitoring of environmental multilateralism. Similarly, to the extent to which domestic state structures move toward the ideal of ecological democracy, one can also be more open to the possibility of domestic environmental policies that go beyond a purely instrumental posture toward the nonhuman world while also taking responsibility to avoid the displacement of social and ecological costs beyond its own territory and into the future.

At present, the European Union represents perhaps the closest real world approximation of such a green Kantian or post-Westphalian culture. However, it is uncertain how far such a culture is likely to spread internationally, where moral persuasion more often takes a back seat to coercion and self-interest in interstate negotiations. In view of the enormous differences between states in terms of military and economic power, infrastructural capacity, culture and organization, and the considerable obstacles in the way of developing a green Kantian or post-Westphalian culture at the international level, it might be argued that the green movement would be better advised to direct its energies toward further reform in the greenest states, in effect "pushing out the green envelope" within these domestic jurisdictions, rather than seeking common ground at the multilateral level (which usually translates into the lowest common denominator). It might be further argued that some of the best environmental policy successes are the result of local experiments and unilateral initiatives at the domestic level, which have led to regional and international "policy diffusion" (i.e., initiatives in some states have served as models or templates that have been voluntarily studied, copied, or adapted by other states rather than from formal multilateral agreements).

Yet the horizontal and vertical diffusion of environmental policy initiatives has rarely been a mutually exclusive development.[5] The greenest states in the world today (mostly found in Northern Europe) also tend to be better international environmental citizens, pushing for stronger environmental treaties and adopting stronger positions (as in the climate change negotiations). Thus mobilizing at the domestic level can have important *multilateral* consequences over time, as a small number of relatively effective green democratic states emerge to take leadership roles in multilateral negotiations, and whose credibility rests in part on their successful domestic environmental initiatives and records. These states sometimes also have an interest in seeking harmonization of their environmental standards to avoid the negative trade effects of higher environmental standards.[6] Finally, preparing for such multilateral meetings can often serve to force the pace domestically and internationally, prompting other parties to consider taking similar initiatives.

A serious question, however, remains: How far can we expect green states to proliferate in world where there is a growing disparity in wealth and capacity within and especially between states? As Andrew Hurrell has argued, "many of the most serious obstacles to sustainability have to do with the domestic weaknesses of particular states and state structures."[7] It is no accident that the processes of ecological modernization have been spearheaded in the developed world. Moreover, while most of the richer states are active shapers of economic globalization, there are many more developing states that are more often aggrieved victims of these processes. These problems are not just the legacy of colonialism but also the result of an international, neoliberal economic order that systematically disadvantages the developing world vis-à-vis the developed world.

There is always reason to hope but little reason to expect that those states sponsoring technical forms of ecological modernization will be detained by the fact that a majority of states are not even in a position to sponsor such a green competitive strategy for their local industries. This state of affairs is unacceptable and represents the most serious challenge to global sustainability. However, both hope and expectations can be raised to the extent to which the economically privileged states pursue deeper, more reflexive strategies of ecological modernization, which in turn presupposes a move toward ecological democracy, since they would necessarily become more preoccupied with both global environmental and economic justice. There are, of course, no encouraging signs that the most powerful states—above all, the United States under the second Bush administration—are moving in this direction. Yet the degree of global interdependence is now such that even superpowers need the cooperation of other states in the longer run. This is the so-called paradox of American power outlined by Joseph Nye, which he argues must lead away from the assertion of "hard power" and toward the practice of "soft power" (including a greater preparedness to act multilaterally).[8] However, this can only be the beginning. Without the deepening of democracy within the most privileged states (and especially the United States), the prospects of structural reform to the international economy, an end to the displacement of environmental problems and the beginning of concerted (as distinct from tokenistic) environmental capacity building in the developing world seem remote. As Robert Paehkle puts it, "Irony of ironies, the route to global governance lies in making the wealthy nations more democratic."[9]

Although I have argued that green public spheres are a condition precedent for the emergence of green democratic states, such states will not materialize or proliferate without political leadership, whether from green parties, social democratic parties, or other social actors. This applies most obviously to elected governments that actively seek to pursue a green agenda, such as the Swedish Social Democratic Party under the leadership of Göran Persson, which embarked in 1996 on "a new and noble mission" to make Sweden an ecologically sustainable society.[10] However, it also applies to other actors in the social, economic, and educational spheres who seek to activate and enhance the state's and society's environmental capacity. Leadership ought not to mean an overweening executive aggressively rushing through a program of reform and ignoring oppositional movements or community know-how and experience. In any event, the constitutional design of the green democratic state should protect citizens from overzealous governments or officials (green or otherwise) while facilitating discursive consensus formation and adaptive policy learning. This includes leaving plenty of room for community initiatives in civil society as well. Nonetheless, in the context of the current order, visionary political leadership is essential for environmental capacity building (including constitutional reform) and the kind of diplomacy that leads to cooperative solutions to common problems.

The welfare state took more than fifty years to emerge; indeed, after another fifty years it is still holding out against the forces of economic globalization, albeit in a weakened form. There are lessons here for those persuaded by the idea of the green democratic state. As James Meadowcroft makes clear in his discussion of the "ecostate," it will be a protracted and conflict-ridden struggle, the green movement will face difficult odds and there are no guarantees.[11] However, if the multifarious green movement is able to maintain critical and vibrant domestic and transnational green public spheres and social movements with a vigorous electoral arm in all tiers of government, working through the party system to influence and ultimately capture conventional political power, then the green democratic state might become a real possibility.

Notes

Chapter 1

1. Throughout this book I use the phrase "environmental movement" to refer to all those citizens' initiatives, movements, and nongovernmental organizations that are specifically concerned to promote environmental protection. The phrase "green movement" is used in a broader sense to encompass all those new social movements and organizations that are concerned to integrate ecological sustainability with social justice. This goal is also reflected in the broad platform that is common to virtually all green parties (ecological responsibility, social justice, grassroots democracy and nonviolence). Many, but not all, environmental movements share these broader goals, which explains why the phrases are sometimes used interchangeably.

2. Hugh Emy and Paul James, Debating the state: The real reasons for bringing the state back in, in Paul James, ed., *The State in Question* (Sydney: Allen and Unwin, 1996), 8–37 at 14.

3. Andrew Vincent, *Theories of the State* (Oxford: Blackwell, 1987), 224.

4. This tradition is explained in section 1.2.

5. Vincent, *Theories of the State*, 224.

6. I enlist the composite term "nation-state" (rather than just "state") here since this leads us to think about states *in relation* to "peoples," understood as a special kind of political community held together by any one or more of a number of different ties, including cultural, religious, and/or linguistic ties and/or shared institutional legacies or even just a common history. Nonetheless, for much of the discussion I will be focusing on the "state" rather than the "nation" because this is where most of the green critique has been directed. Accordingly, since state-building and nation-building are two overlapping developments, I will use the shorthand "state" unless questions of national identity and nationhood arise, such as in chapter 7.

7. The eco-Marxists' analysis is represented by Colin Hay, From crisis to catastrophe? The ecological pathologies of the liberal-democratic state, *Innovations* 9(1996): 421–34; Colin Hay, Environmental security and state legitimacy,

Capitalism, Nature, Socialism, 5(1994): 83–97; James O'Connor, *Natural Causes: Essays in Ecological Marxism* (New York: Guilford Press, 1998). The radical political ecology position is variously represented by Ken Conca, Rethinking the ecology-sovereignty debate, *Millennium* 23(1994): 701–11; Ken Conca, Beyond the statist frame: Environmental politics in a global economy, in Fred P. Gale and R. Michael M'Gonigle, eds., *Nature, Production, Power: Towards an Ecological Political Economy* (Cheltenham, UK: Edward Elgar, 2000), 141–55; John S. Dryzek, Ecology and discursive democracy: Beyond liberal capitalism and the administrative state, *Capitalism, Nature, Socialism* 3(2) (1992): 18–42; Eric Laferrière, Emancipating international relations theory: An ecological perspective, *Millennium* 25(1996): 53–75; Ronnie D. Lipschutz and Ken Conca, eds., *The State and Social Power in Global Environmental Politics* (New York: Columbia University Press, 1993); Matthew Paterson, Green political strategy and the state, in N. Ben Fairweather, Sue Elworthy, Matt Stroh, and Piers H. G. Stephens, eds., *Environmental Futures* (London: Macmillan, 1999), 73–87; Wolfgang Sachs, ed., *Global Ecology* (London: Zed Books, 1993); Julian Saurin, International relations, social ecology and the globalisation of environmental change, in John Vogler and Mark Imber, eds., *The Environment and International Relations* (London: Routledge, 1996), 77–98; Michael Zürn, The rise of international environmental politics, *World Politics* 50 (1998): 617–49. The ecofeminist analysis is represented by Vandana Shiva, *Staying Alive: Women, Ecology and Development* (London: Zed Books, 1988). Prominent ecoanarchists include Murray Bookchin, *The Ecology of Freedom: The Emergence and Dissolution of Hierarchy* (Palo Alto, CA: Cheshire Books, 1982), and Alan Carter, *A Radical Green Political Theory* (London: Routledge, 1999).

8. See, for example, Conca, Beyond the statist frame; Lipschutz and Conca, *The State and Social Power in Global Environmental Politics*; Saurin, International relations, social ecology and the globalisation of environmental change; Zürn, The rise of international environmental politics.

9. This development is broadly in line with the "new sociology." For an introductory overview, see Kate Nash, *Contemporary Political Sociology: Globalization, Politics, and Power* (Malden, MA: Blackwell, 2000).

10. Hedley Bull, The state's positive role in world affairs, *Daedalus* 108(1979): 111–23.

11. Bull, The state's positive role in world affairs, 112.

12. Gianfranco Poggi, *The State: Its Nature, Development and Prospects* (Stanford: Stanford University Press, 1990), 18.

13. Lennart J. Lundqvist, A green fist in a velvet glove: The ecological state and sustainable development, *Environmental Values* 10(4) (2001): 455–72, 457.

14. Max Weber, "Politics as Vocation" in *From Max Weber: Essays in Sociology*, translated and edited with an introduction by H. H. Gerth and C. Wright Mills (London: Routledge and Kegan Paul, 1948), 78.

15. Weber, Politics as vocation, 77–78.

16. Peter Lassman, "Politics, Power and Legitimation" in Stephen Turner, ed., *The Cambridge Companion to Weber* (Cambridge: Cambridge University Press, 2000), 83–98, especially 88 and 97.

17. Poggi, *The State*, vi.

18. The broad tradition of critical theory may be traced to Marx but now encompasses contemporary theories inspired by the political thought of Antonio Gramsci and the Frankfurt School of Social Research. An even broader rendering of critical social theory might include any social and political theory that is critical of the status quo, including those feminist and poststructural approaches (including the work of Michel Foucault) that share a democratic impulse, an emphasis on the social construction of power and knowledge, and a concern to expose or unmask the interests behind the production of power and knowledge. I principally build on the modernist (i.e., neo-Gramscian and neo-Habermasian) rather than postmodernist expressions of critical theory. For excellent overviews of critical theory, see Andrew Linklater, The achievements of critical theory, in Steve Smith, Ken Booth, and Marysia Zalewski, eds., *International Theory: Positivism and Beyond* (Cambridge: Cambridge University Press, 1995), and Richard Wyn Jones, Introduction: Locating critical international relations theory, in Richard Wyn Jones, ed., *Critical Theory and World Politics* (Boulder, CO: Lynne Rienner, 2001), 1–19.

19. Richard Devetak, Critical theory, in Scott Burchill, Richard Devetak, Andrew Linklater, Matthew Paterson, Christian Reus-Smit, and Jacqui True, eds., *Theories of International Relations*, 2nd ed (Basingstoke, England: Palgrave, 2001), 156.

20. Andrew Linklater, *The Transformation of Political Community* (Cambridge: Polity Press, 1998), 5.

21. Max Horkheimer, *The Eclipse of Reason* (New York: Oxford University Press, 1947), 178.

22. Richard Price and Christian Reus-Smit, Dangerous liaisons: Critical international theory and constructivism, *European Journal of International Relations* 4(1998): 259–94. See also Wyn Jones, Introduction.

23. See, for example, Maarten A. Hajer, *The Politics of Environmental Discourse: Ecological Modernization and the Policy Process* (Oxford: Clarendon Press, 1995), for an especially sophisticated treatment of discourse analysis and environmental politics.

24. See Robyn Eckersley, Habermas and green political thought: Two roads diverging, *Theory and Society* 19(1990): 739–76.

25. Major works in this tradition include: John Barry, *Rethinking Green Politics* (London: Sage Publications, 1999); Robert J. Brulle, *Agency, Democracy and Nature: The U.S. Environmental Movement from a Critical Theory Perspective* (Cambridge: MIT Press, 2000); Andrew Dobson, *Green Political Thought*, 3rd ed. (London: Routledge, 2000); John S. Dryzek, *Rational Ecology: Environment and Political Economy* (Oxford: Basil Blackwell, 1987); John S. Dryzek,

Discursive Democracy: Politics, Policy and Political Science (Cambridge: Cambridge University Press, 1990); Tim Hayward, *Ecological Thought: An Introduction* (Cambridge: Polity Press, 1994); Hayward, *Political Theory and Ecological Values* (New York: St Martin's Press, 1998); Mary Mellor, *Feminism and Ecology* (New York: New York University Press, 1997); Douglas Torgerson, *The Promise of Green Politics: Environmentalism and the Public Sphere* (Durham: Duke University Press, 1999); Val Plumwood, *Feminism and the Mastery of Nature* (London: Routledge, 1993); and Val Plumwood, *Environmental Culture: The Ecological Crisis of Reason* (London: Routledge, 2002).

26. Mellor, *Feminism and Ecology*, 190. Mellor borrows the word "parasitism" from Martin O'Connor, On the misadventures of capitalist nature, in Martin O'Connor, ed., *Is Capitalism Sustainable?* (New York: Guilford Press, 1974).

27. Plumwood, *Environmental Culture*, 72–80.

28. As Andrew Hurrell has noted, "On closer inspection many of the anti-statist arguments of the environmental movement turn out to be calls for reformed and more democratic states." See Andrew Hurrell, A crisis of ecological viability? Global environmental change and the nation state, *Political Studies* XLII (1994): 146–65 at 159.

29. Claus Offe, *Modernity and the State* (Cambridge: MIT Press, 1996), 67.

30. Paterson, Green political strategy and the state, 85.

31. Jürgen Habermas, *Between Facts and Norms*, trans. by William Rehg (Cambridge: Polity Press, 1997), 364.

32. Ulrich Beck uses the term "organized irresponsibility" in *Ecological Politics in an Age of Risk* (Cambridge: Polity Press, 1995), 2, 63–65.

Chapter 2

1. Michael Saward, Green state/democratic state, *Contemporary Politics* 4(1998): 345–56, 345.

2. See, for example, Ken Walker, The state in environmental management: The ecological dimension, *Political Studies* 37(1989): 25–39. Charles Tilly, War making and state making as organized crime, in Peter Evans, Dietrich Reuschemeyer, and Theda Skocpol, eds., *Bringing the State Back In* (Cambridge: Cambridge University Press, 1985).

3. Saward, Green state/democratic state, 347.

4. Lamont Hempel, *Environmental Governance: The Global Challenge* (Washington, DC: Island Press, 1996), 153.

5. I adapt and update this question from one posed by Andrew Hurrell and Benedict Kingsbury, eds., *The International Politics of the Environment* (Oxford: Clarendon Press, 1992), 1. The quotation is not strictly verbatim so I have not used quotation marks, although I do follow closely their formulation.

6. In this respect E. H. Carr's critique of liberal idealism in the interwar years remains sobering. See E. H. Carr, *The Twenty Years' Crisis: An Introduction to the Study of International Relations* (London: Macmillan, 1946).

7. Hans J. Morgenthau, *Politics among Nations: The Struggle for Power and Peace*, brief ed., rev. by Kenneth W. Thompson (New York: McGraw Hill, 1993).

8. Steve Smith, The environment on the periphery of international relations, *Environmental Politics*, 2(1993): 28–45.

9. Margaret Thatcher, Speech to Scottish Conservative Party Conference, 14 May 1982, reproduced in A. Partington, ed., *The Oxford Dictionary of Quotations* (4th ed, 1992), 691.

10. Smith, The environment on the periphery of international relations, 41.

11. Garrett Hardin, The tragedy of the commons, *Science* 162(1968): 1243–48; Walker, The state in environmental management; Ronald John Johnston, *Environmental Problems: Nature, Economy and State* (New York: Belhaven Press, 1989), 138–42; Partricia Mishe, Ecological scarcity in an interdependent world, in R. Falk, R. C. Johansen, and S. S. Kim, eds., *The Constitutional Foundations of World Peace* (Albany: State University of New York Press, 1993): 101–25. As Eric Laferrière and Peter Stoett note, even Richard Falk's early analysis of the endangered planet implicitly subscribed to this realist ontology. See Eric Laferriere and Peter J. Stoett, *International Relations Theory and Ecological Thought: Towards a Synthesis* (London: Routledge, 1999), 152 and 182; Richard Falk, *This Endangered Planet: Prospects and Proposals for Human Survival* (New York: Vintage, 1971).

12. See, for example, Michel Frédérick, A realist's conceptual definition of environmental security, in Daniel H. Deudney and Richard A. Matthew, eds., *Contested Grounds: Security and Conflict in the New Environmental Politics* (Albany: State University of New York Press, 1999), 91–108.

13. Richard K. Ashley, The poverty of neorealism, *International Organization* 38(1984): 225–86, 237.

14. Andrew Linklater, The achievements of critical theory, in Steve Smith, Ken Booth, and Marysia Zalewski, eds., *International Theory: Positivism and Beyond* (Cambridge: Cambridge University Press, 1996), 279–300, 283.

15. Matthias Finger, Global environmental degradation and the military, in Jyrki Käkönen, ed., *Green Security or Militarized Environment?* (Aldershot, England: Dartmouth, 1994), 169–91 at 177. The US Rocky Mountain Arsenal, the site of chemical mismanagement for over 30 years and the largest of the seriously contaminated US Defense Department sites, has been called the "most contaminated square mile on earth" (Odelia Funke, Environmental dimensions of national security, in Jyrki Käkönen, ed., *Green Security or Militarized Environment?* (Aldershot, England: Dartmouth, 1994), 55–82 at 61).

16. See, for example, Michael Zürn, The rise of international environmental politics, *World Politics* 50(1998): 617–49; John Vogler, Introduction. The environment in international relations: Legacies and contentions, in John Vogler and

Mark Imber, eds., *The Environment and International Relations* (Routledge, London, 1996), 1–21; and Laferrière and Stoett, *International Relations Theory and Ecological Thought.*

17. Alexander Wendt, *Social Theory of International Politics* (Cambridge: Cambridge University Press, 1999).

18. As Christian Reus-Smit explains, "Societies of states are communities of mutual recognition; they are bound together by intersubjective meanings that define what constitutes a legitimate state and what counts as appropriate state conduct." (Christian Reus-Smit, *The Moral Purpose of the State*, Princeton, Princeton University Press, 1999), 156.

19. See, for example, Michael Taylor, *The Possibility of Cooperation* (Cambridge: Cambridge University Press, 1987); R. J. Axelrod, *The Evolution of Cooperation* (New York: Basic Books, 1984); and Elinor Ostrom, *Governing the Commons: The Evolution of Institutions for Collective Action* (Cambridge University Press: Cambridge, 1990).

20. Stephen Krasner, ed., *International Regimes* (Ithaca: Cornell University Press, 1983), 2.

21. See, for example, Peter M. Haas, Robert O. Keohane, and Mark A. Levy, eds., *Institutions for the Earth: Sources of Effective International Environmental Protection* (Cambridge, MIT Press 1993); Oran R. Young, *International Governance: Protecting the Environment in a Stateless Society* (Ithaca: Cornell University Press, 1994) and Gareth Porter, Janet Welsh Brown, and Pamela S. Chasek, *Global Environmental Politics*, 2nd ed. (Boulder, CO: Westview Press, 2000); Detlef Sprinz and Tapani Vaahtoranta, The interest-based explanation of international environmental policy, *International Organization* 48(1994): 77–105.

22. Andreas Hasenclever, Peter Mayer, and Volker Rittberger, *Theories of International Regimes* (Cambridge: Cambridge University Press, 1997), 4.

23. Ken Conca, Rethinking the ecology-sovereignty debate, *Millennium* 23(1994): 701–11 at 702–703.

24. Robert W. Cox, Social forces, states and world orders: Beyond international relations theory, *Millennium: Journal of International Studies* 10(1981): 126–55 at 130. This was to echo Max Horkheimer's distinction between traditional and critical theory, in Max Horkheimer, *Critical Theory: Selected Essays*, trans. by Mathew J. O'Connell et al. (New York: Seabury Press, 1972), 188–243.

25. Haas, Keohane, and Levy, *Institutions for the Earth*, 7.

26. Nayef Samhat, International regimes as political community, *Millennium* 26(1997): 349–78 at 357.

27. Wendt, *Social Theory of International Politics.*

28. Wendt, *Social Theory of International Politics*, 93.

29. See Robyn Eckersley, Soft law/hard politics and the climate change treaty, in Chris Reus-Smit, ed., *The Politics of International Law* (Cambridge: Cambridge University Press, 2004).

30. The criteria of relative vulnerability, capacity to adjust and costs of adjustments are applied, for example, by Sprinz and Vaahtoranta, in The interest-based explanation of international environmental policy. For a more general critique, see Eckersley, Soft law/hard politics and the climate change treaty.

31. John Ruggie, *Constructing the World Polity: Essays on International Institutionalization* (London: Routledge, 1998), xi.

32. See, in particular, Andrew Linklater, *The Transformation of Political Community: Ethical Foundations of the Post-Westphalian Era* (Cambridge: Polity Press, 1998), especially ch. 5.

33. Robert Cox, Critical political economy, in R. Cox et al., eds., *International Political Economy: Understanding Global Disorder* (Halifax, Nova Scotia: Fernwood Publishing, 1995), 35.

34. See, for example, Emmanual Adler, Seizing the middle ground: Constructivism in world politics, *European Journal of International Relations* 3(1997): 319–63; Margaret Keck and Kathryn Sikkink, *Activists beyond Borders: Transnational Issue Networks in International Politics* (Ithaca: Cornell University Press, 1998); Audie Klotz, *Norms in International Relations: The Struggle Against Apartheid* (Ithaca: Cornell University Press, 1995); Richard Price and Christian Reus-Smit, Dangerous liaisons: Critical international theory and constructivism, *European Journal of International Relations* 4(1998): 259–94; Richard Price, Moral norms in world politics, *Pacific Review* 9(1997): 45–72; and Alexander Wendt, Anarchy is what states make of it: The social construction of power politics, *International Organization* 46(1992): 391–426.

35. In explaining the differences between regulative rules and constitutive rules, John Ruggie uses the contrasting example of traffic rules versus the rules of chess. Traffic rules merely regulate an antecedently existing activity whereas the rules of chess create the very possibility of playing chess. Ruggie, *Constructing the World Polity*, 22.

36. Ian Hurd, Legitimacy and authority in international politics, *International Organisation* 53(1999): 379–408 at 379.

37. Hurd, Legitimacy and authority in international politics, 388.

38. Hurd, Legitimacy and authority in international politics, 388.

39. See, for example, Elinor Ostrom's *Governing the Commons*.

40. P. A. Hall, Policy paradigms, social learning and the state, *Comparative Politics* 25(1993): 275–96 at 292.

41. Hurd, Legitimacy and authority in international politics, 385.

42. Hurd, Legitimacy and authority in international politics, 385–86.

43. Ruggie, *Constructing the World Polity*, 22.

44. Daniel Philpott, *Revolutions in Sovereignty: How Ideas Shaped Modern International Society* (Princeton: Princeton University Press, 2001).

45. Ruggie, *Constructing the World Polity*, 34.

46. Hurd, Legitimacy and authority in international politics.

47. Hurd, Legitimacy and authority in international politics, 392.

48. As Robert Cox notes, "Gramsci took over from Machiavelli the image of power as a centaur: half man, half beast, a necessary combination of consent and coercion." See Robert Cox, Gramsci, hegemony and international relations, in R. Cox with T. J. Sinclair, *Approaches to World Order* (Cambridge: Cambridge University Press, 1996), 127. In emphasing the importance of ideas and culture in maintaining and in some cases *disguising* hegemonic practices, Cox has sometimes been described as a "critical realist." See, for example, Richard Falk, The critical realist tradition and the demystification of power: E. H. Carr, Hedley Bull, and Robert W. Cox, in S. Gill and J. Mittelman, eds., *Innovation and Transformation in International Studies* (Cambridge: Cambridge University Press, 1997), 39–55 at 43.

49. Bruce Cronin, The paradox of hegemony: America's ambiguous relationship with the United Nations, *European Journal of International Relations* 7(2001): 103–30 at 106.

50. G. John Ikenberry and Charles A. Kupchan, Socialization and hegemonic power, *International Organization* 44(1990): 283–315 at 285. Unlike Cronin, however, Ikenberry and Kupchan include both coercion/manipulation as well as socialization as forms of power exercised by hegemons.

51. Joseph S. Nye, *The Paradox of American Power: Why the World's Only Superpower Can't Go It Alone* (New York: Oxford University Press, 2002), 176.

52. Nye also reserves the right for America to act unilaterally in certain circumstances (e.g., when basic security and economic interests are threatened). Nye, *The Paradox of American Power*, especially 154–63.

53. See John Horton, *Political Obligation* (Basingstoke, England: Macmillan, 1992), especially at 158–62.

54. This, at any rate, had been the prevailing view of US Secretary of State Colin Powell—a view that was not shared by all senior members of the administration (notably Defense Secretary Donald Rumsfield and Assistant Secretary of Defense Paul Wolfowitz).

55. Hurd, Legitimacy and authority in international politics, 384–85.

56. Cronin, The paradox of hegemony.

57. David Beetham, *The Legitimation of Power* (Atlantic Highlands, NJ: Humanities Press International, 1991), 39.

58. Beetham, *The Legitimation of Power*, 109.

59. William Rehg, Intractable conflicts and moral objectivity: A dialogical, problem-based approach, *Inquiry* 42(1999): 229–58 and Jürgen Haacke, Theory and praxis in international relations: Habermas, self-reflection, rational argument, *Millennium* 25(1996): 255–89.

60. Thomas Risse, "Let's Argue!": Communicative action in world politics, *International Organisation* 54(2000): 1–39, 15–16.

61. The most sustained attempts to apply Habermasian-type insights to international law have been Friedrich V. Kratochwil, *Rules, Norms, and Decisions* (Cambridge: Cambridge University Press, 1989) and Thomas Franck, *The Power of Legitimacy Among Nations* (New York: Oxford University Press, 1990).

62. See, for example, Robyn Eckersley, Soft law, hard politics and the climate change treaty.

63. Jürgen Habermas, *The Inclusion of the Other: Studies in Political Theory* (Cambridge: MIT Press, 1998), 177; See also Kratochwil, *Rules, Norms, and Decisions* and Franck, *The Power of Legitimacy among Nations.*

64. According to Michael Lisowski, an ABC News poll released in April 2001 showed that 61 percent of Americans supported ratification while only 26 percent opposed it. Michael Lisowski, Playing the two-level game: US President Bush's decision to repudiate the Kyoto Protocol, *Environmental Politics* 11(2002): 101–19 at 114.

65. Wendt, *Social Theory of International Politics*, 298–99.

66. Wendt, *Social Theory of International Politics*, 299.

67. Wendt, *Social Theory of International Politics*, 306.

68. Wendt, *Social Theory of International Politics*, 306.

69. Social learning can give rise to a "we-feeling" in particular security communities. See Karl Deutch, *Political Community at the International Level* (New York: Archon Books, 1970), discussed by Linklater, *The Transformation of Political Community*, 119.

70. Linklater, *The Transformation of Political Community*, 166–68.

71. Linklater, *The Transformation of Political Community*, 167.

72. Linklater, *The Transformation of Political Community*, 167.

73. Linklater, *The Transformation of Political Community*, 169.

74. Wendt, *Social Theory of International Politics*, 43.

75. Wendt, *Social Theory of International Politics*, 43.

76. Significant examples are the Convention on Environmental Impact Assessment in a Transboundary Context (adopted 1991, entered into force on 10 September 1997) and the Aarhus Convention (the UN/ECE Convention on Access to Information, Public Participation in Decision-making and Access to Justice in Environmental Matters—adopted June 1998). These are discussed in more detail in chapter 7.

77. William Rees and M. Wackernagel, Ecological footprints and appropriated carrying capacity, in A. Jansson, M. Hammer, C. Folke, and R. Constanza, eds., *Investing in Natural Capital: The Ecological Economics Approach to Sustainability* (Washington, DC: Island Press, 1994).

78. Juan Martínez-Alier, Environmental justice (local and global), *Capitalism, Nature, Socialism* 8(1997): 91–107.

79. John Gerard Ruggie, Territoriality and beyond: Problematizing modernity in international relations, *International Organisation* 47(1993): 144–73.

80. Ruggie, Territoriality and Beyond, 151.

81. Ruggie, Territoriality and Beyond, 157.

82. Ruggie, Territoriality and Beyond, 165. That is, the gradual "negation of the exclusive territorial form has been the locale in which international sociality throughout the modern era has been embedded," 171.

83. Hedley Bull, *Justice in International Relations: The Hagey Lectures* (Waterloo, Ontario: University of Waterloo, 1984), 14.

Chapter 3

1. According to Ngaire Woods, in political science this is the most debated of all propositions about globalization. See Ngaire Woods, The political economy of globalization, in Ngaire Woods, ed., *The Political Economy of Globalization* (London: Macmillan, 2000), 10. In this chapter I focus primarily on economic globalization, by which I mean the expansion of regional and global markets consequent upon changes in production, trade, investment, and finance. However, many of these changes are inextricably tied to changes in communications technologies, new multilateral initiatives, international organizations, and transnational NGOs.

2. Anthony McGrew, Globalisation and the reconfiguration of political power, Paper Presented to the Civilising the State Conference, Deakin University, Melbourne, 5–6 December, 1999, 7. For one of the most comprehensive and balanced studies on this topic, see David Held, Anthony McGrew, David Goldblatt, and Jonathan Perraton, *Global Transformations: Politics, Economics and Culture* (Stanford: Stanford University Press, 1999).

3. Stephen Gill, Globalisation, market civilisation, and disciplinary neoliberalism, *Millennium: Journal of International Studies* 24(1995): 399–423.

4. James O'Connor, *The Fiscal Crisis of the State* (New York: St Martin's Press, 1973); Jürgen Habermas, *Legitimation Crisis*, trans. by Thomas McCarthy (London: Heinemann, 1973); Claus Offe, The theory of the capitalist state and the problem of policy formation, in Leon Lindberg, ed., *Stress and Contradiction in Modern Capitalism* (Lexington, MA: D.C. Heath, 1975); and Claus Offe, *Contraditions of the Welfare State* (Cambridge: MIT Press, 1984).

5. O'Connor, *The Fiscal Crisis of the State*; Habermas, *Legitimation Crisis*; Offe, The theory of the capitalist state and the problem of policy formation; and *Contraditions of the Welfare State*.

6. Claus Offe and W. von Ronge, Theses on the state, *New German Critique* 6 (1975): 139–47.

7. Clyde W. Barrow, *Critical Theories of the State: Marxist, Neo-Marxist, Post*-Marxist (Madison: University of Wisconsin Press, 1993), 97.

8. Theda Skocpol, *States and Social Revolutions* (Cambridge: Cambridge University Press, 1979), 31; see also Skocpol, Bringing the state back in: Strategies of analysis in current research, in Peter Evans, Dietrich Rueschemeyer, and Theda Skocpol, eds., *Bringing the State Back In* (Cambridge: Cambridge University Press, 1985).

9. Barrow, *Critical Theories of the State*, 96.

10. See, for example, Offe, *Contradictions of the Welfare State*, 37, 132, and Habermas, *Legitimation Crisis*, 49.

11. Offe, *Contradictions of the Welfare State*, 88–89.

12. Despite the widespread enlistment of the term *crisis*, the state nonetheless seems to have survived.

13. Donella H. Meadows, Dennis L. Meadows. Jorgen Randers, and William H. Behrens III, *The Limits to Growth* (New York: Universe Books, 1972).

14. Ecological modernisation, as Albert Weale points out, has shown that the conflict between the accumulation and legitimation imperatives is not as great as was once thought. Albert Weale, *The New Politics of Pollution* (Manchester: Manchester University Press), 89.

15. Foremost in this tradition are John S. Dryzek, Ecology and discursive democracy: Beyond liberal capitalism and the administrative state, *Capitalism, Nature, Socialism* 3(2) (1992); 18–42; John S. Dryzek, *Deliberative Democracy and Beyond: Liberals, Critics, Contestations* (Oxford: Oxford University Press, 2000); James O'Connor, *Natural Causes: Essays in Ecological Marxism* (New York: The Guilford Press, 1997); and Colin Hay, From crisis to catastrophe? The ecological pathologies of the liberal-democratic state, *Innovations* 9(4) (1996): 421–34.

16. O'Connor, *Natural Causes* (particularly the essay entitled "Is sustainable capitalism possible?")

17. O'Connor, *Natural Causes*, Chapter 8.

18. O'Connor, *Natural Causes*, 245.

19. For fuller discussion, O'Connor, *Natural Causes*, 246–47.

20. O'Connor, Natural Causes, 235 and 240.

21. Dryzek, Ecology and discursive democracy, 26.

22. Dryzek, Ecology and discursive democracy, 27. For Dryzek, contemporary environmental policy debates may be understood as different backward and forward movements along different sides of the triangle (more or less state centralization or administrative discretion, more or less liberal democracy, or more or less market-based strategies or privatization), but without any fundamental change to the logic of these three social institutions. From *within* the triangle, ecological problems emerge as yet another contradiction of the welfare state (32).

23. Dryzek, *Deliberative Democracy and Beyond*, 97 and 110. I discuss these claims in more detail in chapter 6.

24. Hay, From crisis to catastrophe?

25. Hay, From crisis to catastrophe? 425.

26. Martin Janicke, *State Failure* (Cambridge: Polity Press, 1990) and Ulrich Beck, *The Risk Society: Towards a New Modernity* (London: Sage Publications, 1992).

27. Habermas, *Legitimation Crisis*, 49; Claus Offe, *Disorganized Capitalism* (Cambridge: MIT Press, 1985), 6.

28. Barrow, *Critical Theories of the State*, 118.

29. Barrow, *Critical Theories of the State*, 120.

30. In particular, new social movements are presented as acting somewhat schizophrenic insofar as they make regulatory demands of the state and therefore contribute to the inexorable "rising cost crisis" (part of O'Connor's economic logic) yet are also potential agents of social and structural transformation toward a resolution of this crisis (part of O'Connor's political logic).

31. Stuart Rosewarne, Marxism, the second contradiction and socialist ecology, *Capitalism, Nature, Socialism* 8(1997):99–120, 112.

32. Rosewarne, Marxism, the second contradiction and socialist ecology, 112.

33. Frederick H. Buttel, Some observations on states, world orders, and the politics of sustainability, *Organisation and Environment* 11(1998): 261–86, 261.

34. O'Connor, *Natural Causes*, ch. 14; John S. Dryzek, *The Politics of the Earth: Environmental Discourses* (Oxford: Oxford University Press, 1997).

35. Barrow, *Critical Theories of the State*, 117.

36. John Ruggie, *Constructing the World Polity: Essays on International Institutionalization* (London: Routledge, 1998), 94. See also Donald Polkinghorne, *Narrative Knowing and the Human Sciences* (Albany: State University of New York Press, 1988), 19–20.

37. Alan Schaiburg, *The Environment* (New York: Oxford University Press, 1980). As Buttel notes, Schnaiburg's analysis is more political than economic and therefore much closer to Gramsci (Buttel, Some observations on states, world orders, and the politics of sustainability, 268).

38. Rosewarne, Marxism, the second contradiction, and socialist ecology, 114.

39. See Robyn Eckersley, Disciplining the market, calling in the state: The politics of economy–environment integration, in Stephen Young, ed., *The Emergence of Ecological Modernisation: Integrating the Environment and the Economy* (London: Routledge, 2000), 233–52.

40. John S. Dryzek, *Discursive Democracy: Politics, Policy and Political Science* (Cambridge: Cambridge University Press, 1990), ch. 4.

41. Ulrich Beck, *Ecological Enlightenment: Essays on the Politics of the Risk Society* (Atlantic Highlands, NJ: Humanities Press, 1995); Peter Christoff, Ecological modernisation, ecological modernities, *Environmental Politics* 5(1996): 476–500; and Arthur Moll, Ecological modernisation and institutional reflexiv-

ity: Environmental reform in the late modern age, *Environmental Politics* 5(1996): 302–323.

42. By "political autonomy" I mean, following Held et al., the capacity of state managers and agencies to articulate and pursue policy goals and preferences (Held, McGrew, Goldblatt, and Perraton, *Global Transformations*, 29).

43. Philip Cerny, Paradoxes of the competition state: The dynamics of political globalisation, *Government and Opposition* 36(1997): 251–74 at 259.

44. Susan Strange, The defective state, *Daedalus* 124 (Spring 1995): 55–74 at 55.

45. Ruggie, *Constructing the World Polity*, ch. 2 (Embedded liberalism and the postwar economic regimes); Philip G. Cerny, The infrastructure of the infrastructure? "Toward embedded financial orthodoxy" in the international political economy, in Ronen P. Palan and Barry Gills, eds., *Transcending the State-Global Divide: A Neo-structuralist Agenda in International Relations* (Boulder, CO: Lynne Reinner, 1994), 223–49; Cerny, Paradoxes of the competition state, 259.

46. Robert Cox, Critical political economy, in Robert Cox et al., eds., *International Political Economy: Understanding Global Disorder* (Halifax, Nova Scotia: Fernwood Publishing, 1995), 39.

47. Gill, Globalisation, market civilisation, and disciplinary neoliberalism, 4.

48. David Osborne and Ted Gaebler, *Reinventing Government* (Reading, MA: Addison-Wesley, 1992).

49. See Saskia Sassen, *Losing Control? Sovereignty in an Age of Globalization* (New York: Columbia University Press, 1996), 28; and James M. Goldgeir and Michael McFaul, A tale of two worlds: Core and periphery in the post–cold-war era, *International Organization* 46 (1992): 467–91.

50. Woods, "The Political Economy of Globalization," 10.

51. Kenichi Ohmae, *The End of the Nation State: The Rise of the Regional Economies* (New York: Harper Collines/Free Press, 1995). Linda Weiss, for example, has argued that the hyperglobalists such as Ohmae have tended "to exaggerate state powers in the past in order to claim feebleness in the present", downplay the diversity of state responses to such pressures and gloss over the active role played by states in economic multilateralism. Linda Weiss, *The Myth of the Powerless State: Governing the Economy in a Global Era* (Cambridge: Polity Press, 1998), 190–91 at 193.

52. The active role of states in this process has been emphasised by Philip Cerny, *The Changing Architecture of the State: Structure, Agency and the Future of the State* (London: Sage, 1990); Paul Hirst and Graham Thompson, *Globalization in Question: The International Economy and the Possibilities of Governance* (Cambridge: Polity Press, 1996); Sassen, *Losing Control?*; and Weiss, *The Myth of the Powerless State.*

53. Sassen, *Losing Control?* 25–26.

54. Paul Hirst, *From Statism to Pluralism* (London: UCL Press, 1997), 230–31.

55. Indeed, Martin Shaw has suggested that it is possible to identify an internationalized, pan-Western state, linked together and constituted by a new "global layer" of state power, manifested in international organizations and multilateral arrangements that are more or less controlled by Western states. This new "global layer" is not only mostly dependent on the Western state; it has also transformed the Western state into "post-imperial, internationalized and democratised forms" but with contradictory tendencies. Indeed, Shaw suggests that we think of "The West as a single state with 'many governments.'" Martin Shaw, *Theory of the Global State: Globality as an Unfinished Revolution* (Cambridge: Cambridge University Press, 2000), 254 and 244.

56. For an excellent critique see, Geoffrey Garratt, Globalization and national autonomy, in Ngaire Woods, ed., *The Political Economy of Globalization* (Basingstoke, England: Macmillan, 2000), 107–46.

57. Ken Conca, The WTO and the undermining of global environmental governance, *Review of International Political Economy* 7(2000): 484–94 at 486.

58. See also Garratt, Globalization and national autonomy.

59. Ronan Palan and Jason Abbott, *State Strategies in the Global Economy* (London: Pinter, 1996), 5.

60. See Jonathan Golub, Introduction and overview, in Jonathan Golub, ed., *Global Competition and EU Environmental Policy* (London: Routledge, 1998), 6.

61. David Wheeler, Racing to the bottom? Foreign investment and air pollution in developing countries, *Journal of Environment and Development* 10(2001): 225–45, 238.

62. Maarten A. Hajer, *The Politics of Environmental Discourse: Ecological Modernization and the Policy Process* (Clarendon Press: Oxford, 1995), 30.

63. Eckersley, Disciplining the market, calling in the state.

64. Christoff, Ecological modernisation, ecological modernities, 490.

65. It is a context where "[d]ecisions have to be taken on the basis of a more or less continuous reflection on the conditions of one's action." Anthony Giddens, *The Consequences of Modernity* (London: Polity Press, 1990), 86.

66. Christoff, Ecological modernisation, ecological modernities, 490.

67. See, for example, Jänicke, *State Failure* and Weale, *The New Politics of Pollution.*

68. World Commission on Environment and Development, *Our Common Future: The Report of the World Commission on Environment and Development* (Oxford: Oxford University Press, 1987).

69. Hajer, *The Politics of Environmental Discourse*, 26 and 14.

70. William Ophuls, Leviathan or oblivion? in Herman E. Daly, ed., *Toward a Steady State Economy* (San Francisco: Freeman, 1973).

71. Arthur A. Mol, Ecological modernisation and institutional reflexivity: Environmental reform in the late modern age, *Environmental Politics* 5(1996): 302–23 at 314.

72. Jonathan Golub, Introduction and overview, 20.

73. Ernst von Weizsacker, Amory B. Lovins, and A. and L. Hunter Lovins, *Factor 4: Doubling Wealth—Halving Resource Use* (Sydney: Allen and Unwin, 1997).

74. Edward Goldsmith, *The Great U-Turn: De-industrialising Society* (Hartland, U.K.: Green Books, 1988).

75. Ingolfur Blühdorn, Ecological modernization and post-ecological politics, in Gert Spaargaren, Arthur P. J. Mol and Frederick H. Buttel, eds., *Environment and Global Modernity* (London: Sage, 2000), 219.

76. Martin Jänicke, Okologische und Politische Modernisierung in Entwickelten Industriegesellschaften, in V. von Prittwitz, ed., *Umweltpolitik als Modernisierungsprozess. Politikwissenschaftelicke Umwelt-forschung und—lehre in der Bundesrepublik* (Opladen: Leske and Budrich, 1993), 15–30 at 18, trans. and quoted in Bluhdorn, Ecological modernization and post-ecological politics, 210.

77. Bluhdorn, Ecological modernization and post-ecological politics, 221.

78. Michael Jacobs, *Environmental Modernisation: The New Labour Agenda* (Fabian Society: London, 1999), 4. However, Jacobs's pamphlet should not be read too literally on this score. That is, it should be understood as an attempt to make a *tactical virtue* out of what he calls "Un-greening the environment," thereby making it uncontroversial, unthreatening, and therefore appealing to New Labour in Britain, and readily digestible within the framework of its own modernizing agenda.

79. Lennart J. Lundqvist, Capacity-building or social construction? Explaining Sweden's shift towards ecological modernisation, *GeoForum* 31 (2000): 21–32, 30.

80. Andrew Dobson, *Green Political Thought*, 3rd ed. (London: Routledge, 2000), 213.

81. Christoff, Ecological modernisation, ecological modernities, 485–90.

82. Jacobs, *Environmental Modernisation*, 22–23.

83. Jacobs, *Environmental Modernisation*, 21; Martin Jänicke and Helmut Weidner, eds., *National Environmental Policies: A Comparative Study of Capacity-Building*, in collaboration with H. Jörgens (Berlin: Springer-Verlag, 1997), 299–300.

84. For a discussion of the short term political challenges associated with such restructuring, see Eckersley, Disciplining the market, calling in the state: The politics of economy–environment integration, especially at 248–50.

85. Dobson, *Green Political Thought*, 211; see also Christoff, Ecological modernisation, ecological modernities, 487.

86. Douglas Torgerson, *The Promise of Green Politics: Environmentalism and the Public Sphere* (Durham: Duke University Press, 1999), 145.

87. Torgerson, *The Promise of Green Politics*, 145.

88. Paul Rutherford, Ecological modernisation and environmental risk, in Éric Darier, ed., *Discourses of the Environment*, Oxford: Blackwell, 1999), 117.

89. Hajer, *The Politics of Environmental Discourse*, 34. See also Christoff, Ecological modernisation, ecological modernities, 483.

90. See, for example, Giddens, *The Consequences of Modernity*; and Scott Lash, Bronislaw Szerszynski and Brian Wynne, eds., *Risk, Environment and Modernity: Towards a New Ecology* (London: Sage Publications, 1996).

91. William Rees and M. Wackernagel, Ecological footprints and appropriated carrying capacity, in A. Jansson, M. Hammer, C. Folke, and R. Constanza, eds., *Investing in Natural Capital: The Ecological Economics Approach to Sustainability* (Washington, DC: Island Press, 1994).

92. Giddens, *The Consequences of Modernity.*

93. Robert C. Paehlke, *Democracy's Dilemma: Environment, Social Equity, and the Global Economy* (Cambridge: MIT Press, 2003), 273.

94. Giddens, *The Consequences of Modernity*, 51; Christoff, Ecological modernisation, ecological modernities, 496.

95. Christoff, Ecological modernisation, ecological modernities, 495. Mol likewise argues that the disembeddedness brought about by the processes of modernisation cannot be simply reversed; it can only ever lead to a process of re-embedding economic practices within the institutions of modernity, but in ways that are more "ecologically rational." See Moll, Ecological modernisation and institutional reflexivity, 306.

96. Ulrich Beck, *Ecological Politics in an Age of Risk*, trans. by Amos Weisz (Cambridge: Polity, 1995).

97. Beck, Politics of risk society, 21.

98. Ulrich Beck, *Ecological Enlightenment: Essays on the Politics of the Risk Society*, trans. by Mark A. Ritter (Atlantic Highlands, NJ: Humanities Press, 1995).

99. James Meadowcroft, From welfare state to ecostate, in John Barry and Robyn Eckersley, eds., *The State and the Global Ecological Crisis* (Cambridge: MIT Press, 2004).

100. P. A. Hall, Policy paradigms, social learning and the state, *Comparative Politics* 25(1993): 275–96, 278; and Norman Vig, Toward common learning: Trends in US and EU environmental policy, Lecture for Summer Symposium on The Innovation of Environmental Policy, Bologna, Italy, 22 July 1997.

101. Vig, Toward common learning, 11.

102. See, for example, Weale, *The New Politics of Pollution*, 31–32; Christoff, Ecological modernisation, ecological modernities.

103. Vig, Toward common learning, 4.

104. Steven Bernstein, Liberal environmentalism and global environmental governance, *Global Environmental Politics* 2(2002): 1–16. For a fuller treatment of this argument, see Bernstein, *The Compromise of Liberal Environmentalism* (New York: Columbia University Press, 2001).

105. OECD, *Capacity Development in Environment* (Paris: OECD, 1994), 8.

106. Jänicke and Weidner, *National Environmental Policies*, 1.

107. Jänicke and Weidner, *National Environmental Policies*, 3 and 15.

108. Blühdorn, Ecological modernization and post-ecological politics, 211.

109. Golub, Introduction and overview, 21.

Chapter 4

1. Francis Fukuyama, *The End of History and the Last Man* (London: Penguin Books, 1992).

2. See, for example, Robert Paehlke, Democracy, bureaucracy and environmentalism, *Environmental Ethics* 10 (1988): 291–308.

3. Douglas Torgerson, *The Promise of Green Politics: Environmentalism and the Public Sphere* (Durham: Duke University Press, 1999). Indeed, Torgerson's notion of the "administrative mind" refers to a particular rationality and configuration of power that spills beyond the state to include "big business" and "big science." A focus on the state alone can therefore deflect attention from the broader policy network of actors, processes, and rationalities that comprise the administrative sphere or mind.

4. Torgerson, *The Promise of Green Politics*, xii.

5. For the latest contribution to this debate, see John Barry and Marcel Wissenburg, eds., *Sustaining Liberal Democracy* (Basingstoke, England: Palgrave, 2001). An exception to this claim is Marcel Wissenburg, *Green Liberalism: The Free and the Green Society* (London: UCL Press, 1998), which I discuss below.

6. Martin Janicke, *State Failure* (Cambridge: Polity Press, 1990), 27.

7. Janicke, *State Failure*, 27.

8. Ulrich Beck, *Ecological Politics in an Age of Risk* (Cambridge: Polity Press, 1995), 2 and 63–65.

9. Theda Skocpol, *States and Social Revolutions* (Cambridge: Cambridge University Press, 1979), 31. See also Theda Skocpol, Bringing the state back in: Strategies of analysis in current research, in Peter Evans, Dietrich Reuschemeyer, and Theda Skocpol, eds., *Bringing the State Back In* (Cambridge: Cambridge University Press, 1985).

10. *States and Social Revolutions*, 29.

11. Alan Carter, *A Radical Green Political Theory* (London: Routledge, 1999), 150.

12. Carter, *A Radical Green Political Theory*, ch. 4.

13. Alan Carter, Towards a green political theory, in Andrew Dobson and Paul Lucardie, eds., *The Politics of Nature: Explorations in Green Political Theory* (London: Routledge, 1993), 39–62; see also Carter, *A Radical Green Political Theory*, 203–207.

14. On green governmentality, see Paul Rutherford, Ecological modernisation and environmental risk, in Éric Darier, ed., *Discourses of the Environment* (Oxford: Blackwell, 1999); on policy professionalism, see Douglas Torgerson, Policy professionalism and the voices of dissent: The case of environmentalism, *Polity* 29 (1997): 358. On the conflict between administrative and ecological rationality, see John S. Dryzek, *Rational Ecology: Environment and Political Economy* (Oxford: Basil Blackwell, 1987).

15. John S. Dryzek, Ecology and discursive democracy: Beyond liberal capitalism and the administrative state, *Capitalism, Nature, Socialism* 3 (1992): 18–42 at 33.

16. Dryzk, *Ecological Rationality*.

17. Paul Hirst, *From Statism to Pluralism* (London: UCL Press, 1997), 12.

18. Beck, *Ecological Politics in an Age of Risk*, 69.

19. Hedley Bull, The state's positive role in world affairs, *Daedalus* 108 (1979): 111–23, 114.

20. Bull, The state's positive role in world affairs, 123.

21. One must also acknowledge that many non-state forms of governance—including para-state functions exercised by many environmental NGOs in the provision of debt relief, aid, welfare, and environmental services—are not publicly accountable. For a critical analysis of the role of NGOs, see Jan Aart Scholte, Global civil society, in Ngaire Woods, ed., *The Political Economy of Globalization* (Basingstoke, England: Macmillan, 2000), 173–201.

22. Gianfranco Poggi. *The State: Its Nature, Development and Prospects* (Stanford: Stanford University Press, 1990), 76–79.

23. This point has also been noted by Andrew Vincent, Liberalism and the environment, *Environmental Values* 7 (1998): 443–59 at 450. An early exception was Mark Sagoff, *The Economy of the Earth* (Cambridge: Cambridge University Press, 1988, ch. 7 (Can environmentalists be liberals?). More recently, the gap is slowly being filled by Avner de-Shalit, Is liberalism environment-friendly? in Roger S. Gottlieb, ed., *The Ecological Community: Environmental Challenges for Philosophy, Politics and Morality* (New York, Routledge, 1997), 82–103; Wissenburg, *Green Liberalism: The Free and the Green Society*; and Brian Barry, Sustainability and intergenerational justice, in Andrew Dobson, ed., *Fairness and Futurity: Essays on Environmental Sustainability and Social Justice* (Oxford: Oxford University Press, 1999), 93–117.

24. See, for example, Terry L. Anderson and Donald R. Leal, *Free Market Environmentalism* (San Francisco: Pacific Research Institute for Public Policy, 1991).

25. Wissenburg, *Green Liberalism.*

26. Andrew Dobson, *Green Political Thought*, 3rd ed. (London: Routledge, 2000), 165.

27. As Herman Daly and John Cobb point out, the price mechanism merely claims to ensure an optimal *allocation* of scarce resources (an efficiency issue—and even this is considerably undermined by the lack of so-called perfect competition in the real world), not an optimal *scale* of resource use (an ecological/ethical issue), nor an optimal *distribution* of resources in local, regional, and global communities (a social/ethical issue). Herman E. Daly and John B. Cobb Jr., *For the Common Good: Redirecting the Economy Toward Community, the Environment and a Sustainable Future* (Boston: Beacon, 1989), 145–46.

28. Robyn Eckersley, Disciplining the market, calling in the state: The politics of economy–environment integration, in Stephen Young, ed., *The Emergence of Ecological Modernisation: Integrating the Environment and the Economy* (London: Routledge, 2000), 233–52.

29. See, for example, James Meadowcroft, Planning for sustainable development: Insights from the literature of political science, *European Journal of Political Research* 31 (1997): 427–54.

30. Tim Hayward, *Ecological Thought: An Introduction* (Cambridge: Polity Press, 1994), 205.

31. Hayward, *Ecological Thought*, 203–204.

32. Alasdair MacIntyre, The privatization of good: An inaugural lecture, *Review of Politics* 52 (1990): 344–61.

33. Val Plumwood, Has democracy failed ecology? An ecofeminist perspective, *Environmental Politics* 4 (1996): 134–68 at 146.

34. Charles Lindblom, *The Intelligence of Democracy: Decision Making Through Mutual Adjustment* (New York: Free Press, 1965). For a succinct argument along these lines, see Robert Goodin, Enfranchising the earth, and its alternatives, *Political Studies*, 44: 835–49.

35. Dryzek, Ecology and discursive democracy.

36. Dryzek, Ecology and discursive democracy, 33.

37. The notion of mobilization of bias comes from Eric E. Schattschneider, *The Semisovereign People: A Realist's View of Democracy in America* (New York: Holt, Rinehart and Winston, 1960).

38. Mancur Oslen, *The Logic of Collective Action* (Cambridge: Harvard University Press, 1965).

39. See, for example, Michael Mackay, Environmental rights and the US system of protection: Why the US environmental protection agency is not a rights-based administrative agency, *Environment and Planning* 26 (1994): 1761–85; and A. Dan Tarlock, Earth and other ethics: The institutional issues, *Tennessee Law Review* 56 (1988): 43–76.

40. Tim Hayward, Constitutional environmental rights: A case for political analysis, *Political Studies* 48 (2000): 558–72. The American Public Trust doctrine is a significant exception here.

41. Gary E. Varner, Environmental law and the eclipse of land as private property, in Frederick Ferré and Peter Hartel, eds., *Ethics and Environmental Policy: Theory Meets Practice* (Athens: University of Georgia Press, 1994), 142–60, 143. See also Robert Goodin, Property rights and preservationist duties, *Inquiry* 33 (1990): 401–32 for a discussion of the trusteeship conception of property.

42. Varner, Environmental law and the eclipse of land as private property, 146.

43. Wissenburg, *Green Liberalism*, 172.

44. Wissenburg, *Green Liberalism*, 123.

45. Wissenburg, *Green Liberalism*, 166.

46. Wissenburg, *Green Liberalism*, 168.

47. James Meadowcroft, From welfare state to ecostate, in John Barry and Robyn Eckersley, eds., *The State and the Global Ecological Crisis* (Cambridge: MIT Press, 2004).

48. Meadowcroft, From welfare state to ecostate.

49. Meadowcroft, From welfare state to ecostate.

50. See, for example, Michael Sandel, *Liberalism and the Limits of Justice* (Cambridge University Press, 1982) and Charles Taylor, *Multiculturalism and the Politics of Recognition* (Princeton: Princeton University Press, 1992).

51. See, for example, Freya Matthews, *The Ecological Self* (London: Routledge, 1991); and Val Plumwood, Nature, self, and gender: Feminism, environmental philosophy, and the critique of rationalism, *Hypatia*, 6 (1991): 3–27.

52. Charles Taylor, Atomism, in Charles Taylor, ed., *Philosophy and the Human Sciences: Philosophical Papers 2* (Cambridge: Cambridge University Press, 1985), 87–210.

53. Bhikhu Parekh, The cultural particularity of liberal democracy, *Political Studies* 40 (Special issue 1992): 160–75 at 161.

54. Stephen Gill, Globalisation, market civilisation, and disciplinary neoliberalism, *Millennium: Journal of International Studies* 24 (1995): 399–423.

55. Theodor Adorno and Max Horkheimer, *The Dialectic of Enlightenment* (New York: Herder, 1972), 3.

56. Tim Hayward, *Ecological Thought: An Introduction* (Cambridge: Polity 1994), 19.

57. J. S. Mill stands out as a significant exception in view of his support for the flourishing of both individual and biological diversity and for a stationary state economy. J. S. Mill, *Principles of Political Economy*, ed. by Donald Winch (Harmondsworth, England: Penguin, 1979), 111–17.

58. However, most of this scrutiny has come from liberal *moral* philosophers seeking to rehabilitate utilitarianism or Kantian ethics to take on board the inter-

ests or rights of animals and future generations (e.g., Peter Singer and Tom Regan) rather than political philosophers. However, the number of contributions to the environmental debate by liberal political philosophers is now growing.

59. John Locke, *Two Treatises of Government*, ed. Peter Laslett (Cambridge: Cambridge University Press, 1988). For an excellent discussion of the ecological implications of Locke's arguments, see Susan Leeson, Philosophic implications of the ecological crisis: The authoritarian challenge to liberalism, *Polity* 11 (1979): 305–306.

60. See Mark Sagoff, *The Economy of the Earth* (New York: Cambridge University Press, 1988), 172.

61. See Paehlke, Democracy, bureaucracy and environmentalism; Torgerson, *The Promise of Green Politics*; and Torgerson, Policy professionalism and the voices of dissent.

Chapter 5

1. See, for example, Jürgen Habermas, *Between Facts and Norms: Contributions to a Discourse Theory of Law and Democracy*, trans. W. Rehg (Cambridge: Polity, 1996); and David Held, *Democracy and the Global Order* (Cambridge: Polity Press, 1995).

2. Robert Goodin, Enfranchising the earth, and its alternatives, *Political Studies* 44 (1996): 835–49.

3. Habermas, *Between Facts and Norms*, 127 (my emphasis).

4. David Held has frequently employed the phrase "overlapping communities of fate" in defending his idea of cosmopolitan democracy as a response to globalisation. See, for example, David Held, Democracy and globalization, in Danielle Archibugi, David Held, and Martin Kohler, eds., *Re-imagining Political Community: Studies in Cosmopolitan Democracy* (Cambridge: Polity 1998), 22; and The changing contours of political community, in Barry Holden, ed., *Global Democracy: Key Debates* (London: Routledge, 2000), 30. I prefer the phrase "community-at-risk" over "community-of-fate" since not all members of the community that are exposed to ecological risks necessarily share the same fate (i.e., suffer the same degree of harm). Moreover there is nothing inevitable or predetermined about the outcomes since the incidence of harm following exposure to ecological risks is always unpredictable. I further distinguish my cosmopolitan approach from Held's in chapter 7. Distinctly green extensions of the principle of affected interests can be found in Andrew Dobson, Representative democracy and the environment, in William M. Lafferty, and James Meadowcroft, eds. *Democracy and the Environment: Problems and Prospects*, (Cheltenham, England: Edward Elgar, 1996), 124–39 and Goodin, Enfranchising the earth, and its alternatives.

5. There is now an extensive literature on this question. See, for example, the contributions to Freya Mathews, ed., *Ecology and Democracy* (London: Frank

Cass, 1996); Brian Doherty and Marius de Geus, eds., *Democracy and Green Political Thought: Sustainability, Rights and Citizenship* (London: Routledge, 1996); John Dryzek, Ecology and discursive democracy: Beyond liberal capitalism and the administrative state, *Capitalism, Nature, Socialism* 3(1992): 18–42; John Dryzek, Green reason: Communicative ethics for the biosphere, *Environmental Ethics* 12 (1990): 195–210; Dobson, Representative democracy and the environment; Goodin, Enfranchising the earth, and its alternatives; Douglas Torgerson, Policy professionalism and the voices of dissent: The case of environmentalism, *Polity* 29 (1997): 345–74; John Barry and Marcel Wissenburg, eds., *Sustaining Liberal Democracy* (Basingstoke, England: Palgrave, 2001); and Ben A. Minteer and Bob Pepperman Taylor, eds., *Democracy and the Claims of Nature: Critical Perspectives for a New Century* (Lanham: Rowman and Littlefield, 2002).

6. John Dryzek, *Rational Ecology: Environment and Political Economy* (Oxford: Basil Blackwell, 1987); John Dryzek, Green reason; and Goodin, Enfranchising the earth, and its alternatives.

7. Although, as we will see in the following chapter, Habermas has recently situated his particular model of discursive democracy *between* liberal democracy and republicanism, it still remains much closer to civic republicanism in spirit and in practice.

8. Hannah Arendt, The crisis in culture, in *Between Past and Future: Six Exercises in Political Thought* (New York: Meridan, 1961).

9. Jürgen Habermas, *Toward a Rational Society: Student Protest, Science and Politics*, trans. Jeremy J. Shapiro (London: Heinemann, 1971).

10. John Dryzek, Transnational democracy, *Journal of Political Philosophy* 7 (1999): 30–51 at 44.

11. John Barry, Sustainability, political judgment and citizenship: Connecting green politics and democracy, in B. Doherty and M. de Geus, eds., *Democracy and Green Political Thought* (London: Routledge, 1996), 116.

12. Habermas's most detailed discussion of the circle of moral considerability may be found in *Justification and Application: Remarks on Discourse Ethics* (Cambridge: Polity, 1993).

13. Robert Goodin, Enfranchising the earth, and its alternatives, 844.

14. Of course, including future generations of humans in the circle of moral considerability will indirectly lead to a greater preparedness to protect nonhuman nature, as Bryan Norton has forcefully argued in *Toward Unity among Environmentalists* (Oxford: Oxford University Press, 1991). I address this argument below.

15. Goodin, Enfranchising the earth, and its alternatives, 843.

16. See, for example, Bryan Norton, Intergenerational equity and environmental decisions: A model using Rawls' veil of ignorance, *Ecological Economics* 1 (1989): 137–59; and Norton, *Toward Unity among Environmentalists*, 226–27.

17. For a reply to Norton's convergence hypothesis, see Andrew Dobson, *Justice and the Environment: Conceptions of Environmental Sustainability and Dimensions of Social Justice* (Oxford: Oxford University Press, 1998), 228, 255–62.

18. See, for example, Kate Soper, *What Is Nature?* (Oxford: Blackwell, 1995).

19. Steven Vogel, *Against Nature: The Concept of Nature in Critical Theory* (Albany: State University of New York Press, 1995), 163. See also Steven Vogel, Habermas and the ethics of nature, in Roger Gottlieb, eds., *The Ecological Community* (New York: Routledge, 1997), 175–92.

20. Don E. Marietta Jr., Reflection and environmental activism, in Don E. Marietta Jr. and Lester Embree, eds., *Environmental Philosophy and Environmental Activism* (Lanham, MD: Rowman and Littlefield, 1995), 79–97.

21. Here I agree with Vogel that Habermas has failed to complete the postempiricist turn in the philosophy of science in maintaining his distinction between nature (as known to us by an objective science) and interaction (the social realm of public discourse). Vogel, Habermas and the ethics of nature.

22. Iris Marion Young, Communication and the other: Beyond deliberative democracy, in Syela Benhabib, ed., *Democracy and Difference: Contesting the Boundaries of the Political* (Princeton: Princeton University Press, 1996); and especially Iris Marion Young, Difference as a resource for democratic communication, in James Bohman and William Rehg, eds., *Deliberative Democracy: Essays on Reason and Politics* (Cambridge: MIT Press, 1997).

23. Anne Phillips, Dealing with difference: A politics of ideas or a politics of presence? in Seyla Benhabib, ed., *Democracy and Difference: Contesting the Boundaries of the Political* (Princeton: Princeton University Press, 1996).

24. Young, Difference as a resource for democratic communication.

25. James Bohman, The coming of age of deliberative democracy, *Journal of Political Philosophy* 6 (1999): 400–25 at 410.

26. John S. Dryzek, Discursive democracy vs. liberal constitutionalism, in Michael Saward, ed., *Democratic Innovation: Deliberation, Representation and Association* (London: Routledge, 2000), 78–89.

27. James Johnson, Arguing for deliberation: Some skeptical considerations, in Jon Elster, ed., *Deliberative Democracy* (Cambridge: Cambridge University Press, 1998), 161–184, 173.

28. Edward Said, Representing the colonized: Anthropology's interlocutors, *Critical Inquiry* 15 (1989): 205–25 and Lynne Sanders, Against deliberation, *Political Theory* 25 (1997): 347–76.

29. See, for example, John Dryzek *Deliberative Democracy and Beyond: Liberals, Critics, Contestations* (Oxford: Oxford University Press, 2000); Jon Elster, *Deliberative Democracy* (Cambridge: Cambridge University Press, 1998); Amy Guttman and Dennis Thompson, *Democracy and Disagreement* (Cambridge: Harvard University Press, 1996); and Habermas *Between Facts and Norms*. For

a recent survey of some of these (and other) works, see Bohman, The coming of age of deliberative democracy.

30. Bohman, The coming of age of deliberative democracy, 422.

31. For Guttman and Thompson, reciprocity, publicity, and accountability "address an aspect of the reason giving process: the kind of reason that should be given, the forum in which they should be given and agents to whom they should be given (Guttman and Thompson, *Democracy and Disagreement*, 52).

32. See also Dennis Thompson, Democratic theory and global society, *Journal of Political Philosophy* 7 (1999): 111–25, 124.

33. Cass Sunstein, Deliberation, democracy and disagreement, in R. Bontekoe and M. Stepaniants, eds., *Justice and Democracy: Cross-Cultural Perspectives* (Honolulu: University of Hawaii Press, 1997), 115.

34. Andrew Light and Eric Katz, eds., *Environmental Pragmatism* (London: Routledge, 1996).

35. I wish to thank Mark Edmondson for first bringing to my attention to this less than fully investigated problem. That is, will deliberation founder if there is no mutual respect between the parties at the beginning of deliberation, or will mutual respect develop during the course, or as a result, of deliberation? What nondeliberative factors might account for the absence or presence of mutual respect before, during or after deliberation?

36. See, for example, Sanders, Against deliberation.

37. Young, Communication and the other; and Young, Difference as a resource for democratic communication.

38. Roberto Gargarella, Full representation, deliberation, and impartiality, in Jon Elster, ed., *Deliberative Democracy* (Cambridge: Cambridge University Press, 1998), 260–80, 261–62.

39. Sunstein has argued that deliberative democracy is necessarily representative "on the theory that direct democracy is less likely to be pervaded by reasons." Sunstein, Deliberation, democracy and disagreement, 94.

40. Phillips, "Dealing with Difference: A Politics of Ideas or a Politics of Presence?"

41. Dennis Thompson, "Democratic Theory and Global Society," 121–122.

42. See Dobson, "Representative Democracy and the Environment."

43. At the international level, the formal inclusion of major international environmental NGOs in multilateral negotiations over environmental treaties and declarations as well as in the governance structures of international financial and trade organisations (e.g., the World Bank, the IMF, and the GATT/WTO) would also provide a more systematic and fulsome representation of affected ecological communities in different environmental policy domains.

44. For a similar argument, see John Barry, Greening liberal democracy: Practice, theory and political economy, in Barry and Wissenburg, eds., *Sustaining Liberal Democracy* (Basingstoke, England: Palgrave, 2001), 59–80.

45. Article 130r of the European Community Treaty, as amended by the Treaty of Union, provides for Community policy to be based on the precautionary principle (along with the polluter pays principle), although no specific obligations are imposed and the implementation of these provisions depend on the recognition of national courts (see Tim Hayward, Constitutional environmental rights: A case for political analysis, *Political Studies* 48(2000): 558–72, 561). For an overview of the situation in Australia, see Charmian Barton, The status of the precautionary principle in Australia: Its emergence in legislation and as a common law doctrine, *Harvard Environmental Law Review* 22(1998): 509–58.

46. Barry, Greening liberal democracy, 71.

47. Fatma Zohra Ksentini, *Final Report of the UN Sub-Commission on Human Rights and the Environment* (UN Doc. E/CN.4/Sub. 2/1994/9, 6 July 1994). For a more extensive discussion, see Robyn Eckersley, Environmental rights and democracy, in Roger Keil, David Bell, Peter Penz, and Leesa Fawcett, eds., *Political Ecology: Global and Local* (London: Routledge, 1998), 353–376 at 362 and note 3.

48. Hayward, Constitutional environmental rights, 558. See also Robyn Eckersley, Liberal democracy and the rights of nature: The struggle for inclusion, in Freya Matthews, ed., *Ecology and Democracy* (London: Frank Cass, 1996), 169–198; and Eckersley, Environmental rights and democracy.

49. Hayward, Constitutional environmental rights, 568. See also James W. Nickel and Eduardo Viola, Integrating environmentalism and human rights, *Environmental Ethics* 16(1994): 265–73.

50. Ronald E. Klipsch, Aspects of a constitutional right to a habitable environment: Towards an environmental due process, *Indiana Law Journal* 49(1974): 203–37 at 229. See also L. Pulido, Restructuring and the contraction and expansion of environmental rights in the United States, *Environment and Planning* A26 (1994): 915–36. According to Pulido, mere procedural justice can be inadequate when minority communities are excluded from private production decisions.

51. Eckersley, Environmental rights and democracy. These procedural environmental rights have also been endorsed by the UN Special Rapporteur on Human Rights and the Environment Fatma Zohra Ksentini. For a detailed discussion, see James Cameron and Ruth Mackenzie, Access to environmental justice and procedural rights in international institutions, in *Human Rights Approaches to Environmental Protection* (Oxford: Clarendon Press, 1996), 129–52 at 133.

52. Michael R. Anderson, Human rights approaches to environmental protection: An overview, in *Human Rights Approaches to Environmental Protection* (Oxford: Clarendon Press, 1996), 9.

Chapter 6

1. Jürgen Habermas, *The Structural Transformation of the Public Sphere: An Inquiry into a Category of Bourgeois Society*, trans. Thomas Burger with the assistance of Frederick Lawrence (Cambridge: MIT Press, 1991).

2. Rattan Gurpreet, Prospects for a contemporary republicanism, *Monist* 84 (2001), 113–30, 118.

3. Habermas, *The Structural Transformation of the Public Sphere*, 140.

4. Habermas, *The Structural Transformation of the Public Sphere*, 235.

5. Habermas, *The Structural Transformation of the Public Sphere*, 195.

6. Jürgen Habermas, *Between Facts and Norms* (Cambridge: Polity Press, 1997), 364. Earlier, Habermas had argued that "The outcome of the struggle between a critical publicity and one that is merely staged for manipulative purposes remains open." Habermas, *The Structural Transformation of the Public Sphere*, 235.

7. Habermas, *The Theory of Communicative Action, Volume Two: Lifeworld and System. A Critique of Functionalist Reason* (Boston: Beacon Press, 1987).

8. These ideas were foreshadowed in *Toward a Rational Society* (London: Heinneman, 1971).

9. *Between Facts and Norms*, 376.

10. *Between Facts and Norms*, 379 and 381.

11. Jürgen Habermas, Human rights and popular sovereignty: The liberal and republican versions, *Ratio Juris* 7(1994): 1–13 at 13.

12. See *Between Facts and Norms*, 296 and 549, note 10, where he points out that Rawls and Dworkin cannot be *confined* to this tradition (my emphasis).

13. Philip Petitt, *Republicanism: A Theory of Freedom and Government* (Oxford University Press, Oxford, 1997).

14. *Between Facts and Norms*, 298.

15. *Between Facts and Norms*, 297–99; and Jürgen Habermas, Three normative models of democracy, in Seyla Benhabib, ed., *Democracy and Difference: Contesting the Boundaries of the Political* (Princeton University Press, 1996), 21–30 at 21.

16. *Between Facts and Norms*, 297. Rousseau's idea that citizens were no longer free once they granted their right of self-governance to representatives is rejected as clearly untenable in large, complex, and plural societies (Habermas, Three normative models of democracy, 29).

17. Habermas, Three normative models of democracy, 26; *Between Facts and Norms*, 298.

18. Habermas, Three normative models of democracy, 29.

19. *Between Facts and Norms*, 308.

20. Jürgen Habermas, *The Postnational Constellation: Political Essays*, translated by Max Pensky (Cambridge: Polity, 2001), 74 and 76. See also Jürgen Habermas, *The Inclusion of the Other: Studies in Political Theory*, ed., Ciaran Cronin and Pablo De Greiff (Cambridge: MIT Press, 1998), lecture 5.

21. According to Habermas, civil society "institutionalizes problem-solving discourses on questions of general interest inside the framework of organized public spheres" (*Between Facts and Norms*, 367). Public spheres are "a communication structure rooted in the lifeworld through the associational networks of civil society" (*Between Facts and Norms*, 361).

22. *Between Facts and Norms*, 360.

23. *Between Facts and Norms*, 370. See Jean L. Cohen and Andrew Arato, *Civil Society and Political Theory* (Cambridge: MIT Press, 1992).

24. *Between Facts and Norms*, 370.

25. *Between Facts and Norms*, 368.

26. *Between Facts and Norms*, 373.

27. *Between Facts and Norms*, 306.

28. *Between Facts and Norms*, 111.

29. Jürgen Habermas, *Justification and Application: Remarks on Discourse Ethics*, trans. Ciaran Cronin (Polity, Cambridge, 1993), 10.

30. *Between Facts and Norms*, 157.

31. *Between Facts and Norms*, 153. As Habermas explains: "Legal norms must assume the form of comprehensible, consistent, and precise specifications, which normally are formulated in writing; they must be made known to all addressees, hence be public; they may not claim retroactive validity; and they must regulate the given set of circumstances or 'fact situation' in terms of general features and connect these with legal consequences in such a way that they can be applied to all persons and all comparable cases in the same way" (*Between Facts and Norms*, 143–44).

32. *Between Facts and Norms*, 155.

33. According to Habermas, "compromises provide for an arrangement that (a) is more advantageous to all than no arrangement whatsoever, (b) excludes free riders who withdraw from cooperation, and (c) excludes exploited parties who contribute more to the cooperative effort than they gain from it" (*Between Facts and Norms*, 166). Whereas a consensus rests on particular reasons that convince all parties, a compromise is an agreement that parties find convincing for *different* reasons.

34. *Between Facts and Norms*, 166. However, Habermas explains that such procedures, which must themselves be justified in moral discourses, would still need to ensure the equal consideration of the interests of all parties (p. 167).

35. *Between Facts and Norms*, 307.

36. *Between Facts and Norms*, 371–72.

37. According to Habermas, the processes of globalization have forced nation-states to open up internally at the same time as they have shrunk the capacity for effective social steering by states (*Postnational Constellation*, 84). For Habermas, the problem is made worse by the fact that many domestic problems can

only be effectively solved at the supranational level, but that supranational governance cannot replicate what welfare states (in their heyday?) were able to achieve (e.g., sufficient levels of social solidarity to enable taxation and redistribution). While Habermas entertains the future possibility of a postnational democracy developing at the pan-European level, he considers the prospects for such a development at the global level to be exceedingly remote.

38. *Between Facts and Norms*, 376 and 381.

39. *Between Facts and Norms*, 145.

40. John Dryzek, *Deliberative Democracy and Beyond: Liberals, Critics, Contestations* (Oxford: Oxford University Press, 2000), 26.

41. Dryzek, *Deliberative Democracy and Beyond*, 26.

42. Whether the essence of democratic republicanism is understood to be the cultivation of public virtue or the minimization of the exercise of arbitrary power, as constitutional designers, republicans have always been masters of suspicion, not naïve idealists. See John Uhr, Instituting republicanism: Parliamentary vices, republican virtues, *Australian Journal of Political Science* 28, Special issue, 27–39 at 29.

43. This point is well made by Pettit, *Republicanism*, 226.

44. Douglas Torgerson, *The Promise of Green Politics: Environmentalism and the Public Sphere* (Durham: Duke University Press, 1999); Robert J. Brulle, *Agency, Democracy and Nature: The U.S. Environmental Movement from a Critical Theory Perspective* (Cambridge: MIT Press, 2000); and Dryzek, *Deliberative Democracy and Beyond.*

45. Torgerson, *The Promise of Green Politics,* particularly 20 and 158.

46. *The Promise of Green Politics*, 20. Torgerson's point is not to argue against goal-directed activity but rather to point to the disunity of theory and practice, and to highlight the tensions in the metaphors and practices of "spheres" and "movements."

47. Joshua Cohen, Procedure and substance in deliberative democracy, in James Bohman and William Rehg, eds., *Deliberative Democracy: Essays on Reason and Politics* (Cambridge: MIT Press, 1997), 426–27.

48. Robyn Eckersley, Environmental pragmatism, ecocentrism and deliberative democracy: Between problem-solving and fundamental critique, in Bob Pepperman Taylor and Ben Minteer, eds., *Democracy and the Claims of Nature* (Lanham: Rowman and Littlefield, 2002), 49–69.

49. Jürgen Haacke, Theory and praxis in international relations: Habermas, self-reflection, rational argument, *Millennium* 25(1996): 255–89 at 279.

50. Ernesto Laclau and Chantal Mouffe, *Hegemony and Socialist Strategy: Towards a Radical Democratic Politics* (London: Verso, 1985); and James O'Connor, *Natural Causes: Essays in Ecological Marxism* (New York: Guilford Press, 1997).

51. Torgerson, *The Promise of Green Politics*, 48.

52. Torgerson is not suggesting that we abandon all goal-directed action. Rather, his concern is to highlight the tensions in the metaphors and practices of spheres and movements. See also Douglas Torgerson, Farewell to the green movement? Political action and the green public sphere, *Environmental Politics* 9(2000): 1–19.

53. Dryzek, *Deliberative Democracy and Beyond*, 23.

54. Iris Marion Young, *Democracy and Inclusion* (Oxford: Oxford University Press, 2000), 160.

55. Young, *Democracy and Inclusion*, 189. Young goes on to distinguish different levels of civil society associations (private, civic, and overtly political), noting movement backwards and forwards between these different levels (p. 163).

56. This is not to ignore the increasing number of campaigns directed at corporations.

57. Young, *Democracy and Inclusion*, 156.

58. Young, *Democracy and Inclusion*, 190.

59. Young, *Democracy and Inclusion*, 156.

60. Young, *Democracy and Inclusion*, 186.

61. Young argues that disaffection with the state as a vehicle for justice is misplaced. For example, she criticizes Paul Hirst's case for "associative democracy," according to which civil society would take over some of the functions currently performed by the state and the market in order to expand the realm of self-governance in civil society. Yet she also notes that well-meaning attempts to link civil society more formally into the processes of policy making, implementation and/or evaluation always carry the risk of disciplining citizens rather than the state (*Democracy and Inclusion*, 195).

62. *Democracy and Inclusion*, 173.

63. Of course, the resolution of environmental problems requires much more than the fulfilment of these functions by the state, as I argue throughout this book.

64. Dryzek, *Deliberative Democracy and Beyond*, 4.

65. Dryzek, *Deliberative Democracy and Beyond*, 81.

66. Dryzek, *Deliberative Democracy and Beyond*, 83.

67. Dryzek, *Deliberative Democracy and Beyond*, 85. Dryzek's criteria for further democratization occur across three dimensions: franchise (how inclusive of the parties affected), scope (how inclusive of issues and areas hitherto shielded from democratic deliberation), and authenticity (how real as distinct from symbolic or perfunctory is political participation).

68. Dryzek, *Deliberative Democracy and Beyond*, 109.

69. Dryzek, *Deliberative Democracy and Beyond*, 88.

70. Dryzek, *Deliberative Democracy and Beyond*, 114.

71. Dryzek, *Deliberative Democracy and Beyond*, 98.

72. Bryan Norton, Integration or reduction: Two approaches to environmetnal values, in Andrew Light and Eric Katz, eds., *Environmental Pragmatism* (London: Routledge, 1996), 105–138 at 124.

Chapter 7

1. John Dryzek, Transnational democracy, *Journal of Political Philosophy* 7 (1999): 30–51.

2. *Collins English Dictionary* (London: Collins, 1979), 394.

3. Michael Walzer, *Thick and Thin: Moral Argument at Home and Abroad* (Northbend, IN: University of Notre Dame Press, 1994).

4. Zenon Bankowski and Emilios Christodoulidis, Citizenship bound and citizenship unbound, in Kimberly Hutchings and Roland Dannreuther, eds., *Cosmopolitan Citizenship* (Basingstoke, England: Macmillan, 1999), 83–104.

5. Habermas refers to his approach as "communicative communitarianism," in *The Inclusion of the Other: Studies in Political Theory* (Cambridge: MIT Press, 1998), 139.

6. David Held, Democracy and the global system, in David Held, ed., *Political Theory Today* (Cambridge: Polity Press, 1991), 197–235; David Held, *Democracy and the Global Order* (Cambridge: Polity Press, 1995); David Held, Democracy and globalization, in Daniele Archibugi, David Held, and Martin Kohler, eds., *Reimagining Political Community: Studies in Cosmopolitan Democracy* (Cambridge: Polity Press, 1998); David Held, The transformation of political community: Rethinking democracy in the context of globalization, in Ian Shapiro and Casiana Hacker-Cordón, eds., *Democracy's Edges* (Cambridge: Cambridge University Press, 1999), 84–111; David Held, The changing contours of political community: Rethinking democracy in the context of globalization, in Barry Holden, ed., *Global Democracy: Key Debates* (London: Routledge, 2000); and Daniele Archibugi and David Held, *Cosmopolitan Democracy* (Cambridge: Polity Press, 1995). See also Thomas Pogge, Cosmopolitanism and sovereignty, *Ethics* 103 (1992): 48–75.

7. *The Postnational Constellation: Political Essays* (Cambridge: Polity Press, 2001). See also Jürgen Habermas, A constitution for Europe? *New Left Review* (second series) (2001): 5–26.

8. *Postnational Constellation*, 100.

9. *Postnational Constellation*, 100.

10. But, as Habermas suggests, the possibility that "the sons, daughters, and grandchildren of a barbaric nationalism" might develop into a pan-European people remains so long as we understand the "people" to be constituted by a civic solidarity and transnational trust rather than an ethnic or linguistic one (*Postnational Constellation*, 100).

11. *Postnational Constellation*, 64.

12. *Postnational Constellation*, 64–65.

13. *Postnational Constellation*, 65.

14. See also Richard Devetak and Richard Higgot, Justice unbound? Globalisation, states and the transformation of the social bond, *International Affairs* 75 (1999): 483–98.

15. *Inclusion of the Other*, 139–40.

16. *Inclusion of the Other*, 139 (original emphasis).

17. See David Miller, *On Nationality* (Oxford: Oxford University Press, 1995); David Miller, Bounded citizenship, in Kimberly Hutchings and Roland Dannreuther, eds., *Cosmopolitan Citizenship* (Basingstoke, England: Macmillan, 1999), 62–65; and David Miller In defence of nationality, in John Hutchinson and Anthony D. Smith, eds., *Nationalism: Critical Concepts in Political Science* (London: Routledge, 2000), 1676–94.

18. Miller understands national communities as communities "constituted by mutual belief, extended in history, active in character, connected to a particular territory, and thought to be marked off from other communities by its members' distinct traits." This, according to Miller, is the right kind of boundedness to ground social obligation, mutual trust and hence political self-determination (Miller, In defence of nationality, 1683).

19. Miller, In defence of nationality, 1687.

20. Walzer, *Thick and Thin.*

21. For Kymlicka, vernacular communities provide the primary forum not only for democratic participation in the world today but also for the legitimation of other levels of government (federal, international). This is why politics that transcends the vernacular (e.g., that which takes place in Brussels in the EU or internationally) is invariably elite-dominated and why mass opinion on the issue of further European integration is usually considerably opposed to elite opinion. See Will Kymlicka, *Politics in the Vernacular: Nationalism, Multiculaturalism, and Citizenship* (Oxford: Oxford University Press, 2001), 121–22. See also Charles Taylor, who has argued that the political recognition of the worth of different cultural communities within a broader polity is a prerequisite to a successful democratic politics in multicultural polities (Charles Taylor, *Multiculturalism and the Politics of Recognition* (Princeton: Princeton University Press, 1992).

22. *Politics in the Vernacular*, 124.

23. Held, The transformation of political community, 106. See also Pogge, Cosmopolitanism and sovereignty.

24. Held, The transformation of political community, 105.

25. David Held, Anthony McGrew, David Goldblatt, and Jonathan Perraton, *Global Transformations* (Stanford: Stanford University Press, 1999), 28.

26. Held, *Democracy and the Global Order*, 234.

27. Held, *Democracy and the Global Order*, 234–35.

28. Held, *Democracy and the Global Order*, 235–36.

29. Held, *Democracy and the Global Order*, 237, n. 6.

30. Held, *Democracy and the Global Order*, chapter 12, especially 279–80.

31. Major events such as participation in the Olympic games likewise can foster identification with fellow nationals who compete in or support the games, but reflected glory does not typically translate into civic solidarity of the kind that might ground redistribution within the polity (except perhaps into scholarships for would-be athletes?).

32. Charles Beitz, *Political Theory and International Relations*, 2nd ed. (Princeton: Princeton University Press, 1999) and Thomas Pogge, *Realizing Rawls* (Ithaca: Cornell University Press, 1989), especially part III.

33. Alexander Wendt, A comment on Held's cosmopolitanism, in Ian Shapiro and Casiana Hacker-Cordón, eds., *Democracy's Edges* (Cambridge: Cambridge University Press, 1999), 127–33 at 131.

34. Michael Saward, A critique of Held, in Barry Holden, ed., *Global Democracy: Key Debates* (London: Routledge, 2000), 37–38.

35. Held, *Democracy and the Global Order*, 231–38.

36. *Inclusion of the Other*, 1998, 150.

37. Saward. A critique of Held, 37–38.

38. Danilo Zolo, The lords of peace: From the holy alliance to the new international criminal tribunals, in Barry Holden, ed., *Global Democracy: Key Debates* (London: Routledge, 2000), 73–86.

39. Andrew Linklater, *The Transformation of Political Community* (Cambridge: Polity Press, 1998), 46–47.

40. Linklater, *The Transformation of Political Community,* 47.

41. Linklater, *The Transformation of Political Community,* 49; see also Gerard Delanty, *Citizenship in a Global Age: Society, Culture, Politics* (Houston, TX: Open University Press, 2000), 144–45.

42. See also Dryzek, Transnational democracy. Like Linklater, Dryzek focuses on transboundary discourses and public spheres, not on how *formal* democratic procedures might be transnationalised.

43. Saward. A critique of Held, 35.

44. Extracted from UN/ECE Web page, *http://www.unece.org/env/pp/* (accessed 2 April 2002).

45. Tim Hayward, Greening the constitutional state: Environmental rights in the European Union, in John Barry and Robyn Eckersley, eds., *The State and the Global Ecological Crisis* (Cambridge: MIT Press, 2004).

46. Hayward, Greening the constitutional state, 22.

47. Hayward, Environmental rights in Europe, 22.

48. James Fishkin, *The Voice of the People: Public Opinion and Democracy* (New Haven: Yale University Press, 1995).

49. Saward, A critique of Held.

50. Saward, A critique of Held, 37.

51. Dennis Thompson, Democratic theory and global society, *Journal of Political Philosophy* 7 (1999): 111–25, 121–22.

52. Daniel Deudney, Global village sovereignty: Intergenerational sovereign publics, federal republican earth constitutions, and planetary identities, in Karen T. Litfin, ed., *The Greening of Sovereignty in World Politics* (Cambridge: MIT Press, 1998).

53. Many states contain aspirational statements about the environmental protection of territory in their constitutions, but I am not aware of any such aspirations that extend beyond the nation or the territorial boundaries of the state.

54. Ulrich Beck, *What Is Globalization?* (Cambridge: Polity Press, 2000), 133.

55. Andrew Linklater, Citizenship and sovereignty in the post-Westphalian State, *European Journal of International Relations* 2 (1996): 77–103.

56. Gianfranco Poggi, *The State: Its Nature, Development and Prospects* (Stanford: Stanford University Press, 1990), 177.

57. Paul Hirst, *From Statism to Pluralism* (London: UCL Press, 1997), 11. I am grateful to Ruth Phillips for drawing my attention to this passage.

58. Richard Bellamy and R. J. Barry Jones, Globalisation and democracy—An afterword, in Barry Holden, ed., *Global Democracy: Key Debates* (London: Routledge, 2000), 202–16 at 211.

59. Wendt, A comment on Held's cosmopolitanism, 128–30.

60. Wendt, A comment on Held's cosmopolitanism, 131.

61. Wendt, A comment on Held's cosmopolitanism, 129–30 and 133.

62. In a similar vein, Bellamy and Jones suggest that efforts to supplement the present interstate system of governance seem more promising than schemes to subvert or supplant it. See Bellamy and Jones, Globalisation and democracy—An afterword, 212.

Chapter 8

1. See, for example, Karen T. Litfin, *The Greening of Sovereignty in World Politics* (Cambridge: MIT Press, 1998).

2. Daniel Philpott, *Revolutions in Sovereignty: How Ideas Shaped Modern International Society* (Princeton: Princeton University Press, 2001), 15–21.

3. It has been argued that changes in the rights and responsibilities that attach to statehood do not amount to changes in the *constitutive* rules of sovereignty and that they only amount to changes in the *regulative* rules. According to Georg Sorensen, the constitutive rule content of sovereignty is the constitutional

independence of states (which have territory, people and government); this is a legal, absolute, and unitary condition that has remained unchanged since the seventeenth century. See Georg Sorensen, Sovereignty: Change and continuity in a fundamental institution, *Political Studies* 47 (1999): 590–604, 593–92. According to Sorensen, regulative rules, in contrast, "regulate interaction between the antecedently existing entities that are sovereign states"; they are like traffic rules that regulate the antecedently existing activity of driving (p. 595). While regulative rules of sovereignty have evolved since the seventeenth century (and here Sorensen cites as examples rules that Philpott has included in his three faces of sovereignty: rules of admission, standards of admission and rights, and responsibilities of membership), the constitutive rules have remained fundamentally the same. Against, Sorenson I would argue that changes in Philpott's three faces of sovereignty *are* constitutive changes, not merely regulative changes, because they reconstitute the roles and purposes of the units of political authority—states. In this respect they are more like the rules of a chess game (which constitute the roles and legitimate moves of the pieces in a defined field of play, which have no meaning outside the game of chess) than traffic rules (which merely regulate, but do not constitute the role and meaning of, cars).

4. Philpott, *Revolutions in Sovereignty*, 5.

5. J. Samuel Barkun and Bruce Cronin, The state and the nation: Changing norms and the rules of sovereignty in international relations, *International Organization* 48 (1994): 107–30, 111.

6. Robert H. Jackson, *Quasi-States: Sovereignty, International Relations and the Third World* (Cambridge: Cambridge University Press, 1993).

7. Christian Reus-Smit, Human rights and the social construction of sovereignty, *Review of International Studies* 27 (2001): 519–38.

8. Adil Najam, An environmental negotiation strategy for the South, *International Environmental Affairs* 7 (1995): 249–87. See also James W. Nickel and Eduardo Viola, Integrating environmentalism and human rights, *Environmental Ethics* 16 (1994): 265–73.

9. International Union for the Conservation of Nature and Natural Resources (IUSN), *World Conservation Strategy: Living Resource Conservation for Sustainable Development*, prepared by the IUCN, with United Nations Environment Programme (UNEP), and the World Wildlife Fund (WWF) (Morges, Switzerland: IUCN, 1980).

10. Margaret E. Keck and Kathryn Sikkink, *Activists Beyond Borders: Advocacy Networks in International Politics* (Ithaca: Cornell University Press, 1998), 36.

11. Christian Reus-Smit, The normative structure of international society, in Fen Osler Hampson and Judith Reppy, eds., *Earthly Goods: Environmental Change and Social Justice* (Ithaca: Cornell University Press, 1996), 96–121, 103.

12. These levels build upon P. A. Hall, Policy paradigms, social learning and the state, *Comparative Politics* 25 (1993): 275–96; and Norma Vig, Toward

common learning: Trends in US and EU environmental policy, Presented before the Summer Symposium on Innovation of Environmental Policy, Bologna, Italy, 22 July 1997.

13. Reus-Smit Reus-Smit, The normative structure of international society, 103.

14. One recent effort by Peter Brown argues that states should become trustees for "the commonwealth of life." See also Peter G. Brown, *Ethics, Economics and International Relations: Transparent Sovereignty in the Commonwealth of Life* (Edinburgh: Edinburgh University Press, 2000). It is also noteworthy that the Commission on Global Governance has recommended reviving the United Nations Trusteeship Council and making it responsible, on behalf of all states, for overseeing the protection of the global commons and serving as the chief forum on environmental and related matters. The Commission on Global Governance, *Our Global Neighbourhood: The Report of the Commission on Global Governance* (Oxford: Oxford University Press, 1995), 251–53.

15. Litfin, *The Greening of Sovereignty*, 8–9.

16. Stephen Krasner, *Sovereignty: Organised Hypocrisy* (Princeton: Princeton University Press, 1999).

17. Edith Brown Weiss, International environmental law: Contemporary issues and the emergence of a new world order, *Georgetown Law Journal* 81 (1993): 675–710 at 675.

18. Edith Brown Weiss, International environmental law, 695.

19. See, for example, Dinah Shelton, Human rights, environmental rights and the right to environment, *Stanford Journal of International Law* 28 (1991): 103–38; Nickel and Viola, Integrating environmentalism and human rights; Alan Boyle and Michael R. Anderson, eds., *Human Rights Approaches to Environmental Protection* (Oxford: Clarendon Press, 1996), 1–23; and Tim Hayward, Constitutional environmental rights: A case for political analysis, *Political Studies* 48 (2000): 558–72.

20. Marc Pallemaerts, International environmental law from Stockholm to Rio: Back to the future? in Philippe Sands, ed., *Greening International Law* (London: Earthscan, 1993), 2; Patricia Birnie, Environment protection and development, *Melbourne University Law Review* 20 (1995–96): 66–100 at 84.

21. Martin Lau, Islam and judicial activism: Public interest litigation and environmental protection in the Islamic Republic of Pakistan, in Alan E. Boyle and Michael R. Anderson, eds., *Human Rights Approaches to Environmental Protection*, Oxford: Clarendon Press, 1996), 285–302 at 298.

22. Albert Weale, *The New Politics of Pollution* (Manchester: Manchester University Press, 1992), 79–82.

23. See, for example, Peter Newell, *Climate for Change: Non-state Actors and the Global Politics of Greenhouse* (Cambridge: Cambridge University Press, 2000); and Keck and Sikkink, *Activists beyond Borders*.

24. Sebastian Oberthür and Hermann E. Ott, *The Kyoto Protocol: International Climate Policy for the 21st Century* (Berlin: Springer-Verlag, 1999), 83–84.

25. Rodger Payne, Habermas, discourse norms, and the prospects for global deliberation, Presented before the Annual Meeting of International Studies Association, Los Angeles, 15–18 March 2000.

26. To cite one significant example, according to Michael Grubb, environmental NGOs played a significant role in the steps leading to the Berlin Mandate, in helping to persuade India and Brazil to break the impasse within the G77 in the climate negotiations. See Michael Grubb, with Christiaan Vrolijk and Duncan Brack, *The Kyoto Protocol: A Guide and Assessment* (London: Royal Institute of International Affairs and Earthscan, 1999), 51–52.

27. Article 4.1.i of the United Nations Framework Convention on Climate Change includes a commitment on the part of the Parties to "encourage the widest participation in this process, including that of non-governmental organizations. . . ." *http://unfccc.int/resource/conv/*

28. *Trail Smelter* Case (United States *v.* Canada) 3 R. Int. Arbitration Awards 1905.

29. *Corfu Channel* Case (Merits) (United Kingdom *v.* Albania) (1949) ICJ Reports 4.

30. For an exhaustive and critical examination of this principle, see Prue Taylor, *An Ecological Approach to International Law: Responding to Challenges of Climate Change* (London: Routledge, 1998), 61–143.

31. *Corfu Channel* case and *Lac Lanoux* Arbitration (Spain *v.* France) 12 R. Int. Arbitration Awards 281.

32. Pierre Dupuy, Due diligence in the international law of liability, in *OECD, Legal Aspects of Transfrontier Pollution* (Paris: OECD, 1997), 369–79 at 369, quoted in Prue Taylor, *An Ecological Approach to International Law,* 91; see also Franz Xavier Perrez, The relationship between "permanent sovereignty" and the obligation not to cause transboundary environmental damage, *Environmental Law* 26 (1996): 1187–1212 at 1199.

33. Nuclear Tests Case (Australia *v.* France) [1973] *ICJ Reports* 99; (New Zealand *v.* France) [1974] *ICJ Reports* 135. As it turned out, France voluntarily agreed to halt nuclear testing so the ICJ was not required to make a determination on the specific claim. See Taylor, *An Ecological Approach to International Law*, 72 and 93.

34. Perrez, The relationship between "permanent sovereignty" and the obligation not to cause transboundary environmental damage, 1207.

35. A/res/27/2996, 15 December 1972. Taylor, *An Ecological Approach to International Law*, 75; Brown Weiss, International environmental law, 78; Pallemaerts, International environmental law from Stockholm to Rio, 2.

36. Charter of Economic Rights and Duties of States, Article 30; 1979 Geneva Convention on Long Range Transboundary Air Pollution; the 1982 UN Convention on the Law of the Sea; the Vienna Convention for the Protection of the Ozone Layer; and 1992 The United Nations Framework Convention on Climate Change.

37. Mention should also be made of the 1982 World Charter for Nature which is limited to the conservation and use of living and natural resources, but is more radical, more limited in scope, and much less influential politically than the Stockholm and Rio Declarations. It has had nothing like the scope and impact of the Stockholm and Rio Declarations.

38. Pallemaerts, International environmental law from Stockholm to Rio.

39. Note that the Brundtland Report had argued for the consolidation of international environmental law and had authorized a group of eminent legal experts to draft a set of legal principles for submission to the UN General Assembly (see J. G. Lammers, ed., *Environmental Protection and Sustainable Development: Legal Principals and Recommendations* (London: Graham and Trotman, 1987.) However, these principles have not received the endorsement of the UNGA. It is noteworthy that one of the stated aims of the UNCED conference was to "promot[e] the further development of international environmental law," including an examination of "the feasibility of elaborating general rights and obligations of States . . . in the field of the environment." (GA Resolution 44/228, 22 December 1989, Doc A/RES/44/228, para 15(d) (1989).)

40. GA resolution 626 (VII), 21 December 1952.

41. Article 1, GA resolution 1803 (XVII), 14 December 1962.

42. GA resolution 3201, 9 May 1974. Article 4(c) of the NIEO, after declaring state sovereignty over "natural resources and all economic activities," goes on to declare that, "[i]n order to safeguard these resources," each state has the right to nationalize or transfer ownership to its nationals of such assets (although Article 1(q) of the NIEO contained only one concession toward the environment in affirming "the need for all states to put an end to the waste of natural resources . . ."). The Charter of Economic Rights and Duties of States was also primarily concerned to allow states to regulate foreign investment and the activities of multinational corporations within their territory, including to nationalize assets in the hands of foreign owners.

43. GA resolution 3281, 12 December 1974. The right of permanent sovereignty over natural resources had also appeared early in the drafts of the two international covenants on human rights and eventually appeared in Article I of both the International Covenant on Economic, Social and Cultural Rights and the International Covenant on Civil and Political Rights 1966. Article I recognizes the collective right of all peoples to economic and political self-determination, including the right freely to use, exploit, and dispose of their natural wealth and resources.

44. Nico Schrijver, *Sovereignty Over Natural Resources: Balancing Rights and Duties* (Cambridge: Cambridge University Press, 1997), 369–70.

45. Kamal Hossain, Introduction, in Kamal Hossain and Subrata Roy Chowdhury, eds., *Permanent Sovereignty Over Natural Resources in International Law: Principles and Practice* (New York: St Martin's Press, 1984), x. See also Schrijver, *Sovereignty over Natural Resources.*

46. Najam, An environmental negotiation strategy for the South.

47. Henry Shue, Ethics, the environment and the changing international order, *International Affairs* 71 (1995): 453–61.

48. Shue, Ethics, the environment and the changing international order, 457.

49. Michael Barnett, The new United Nations politics of peace: From juridical sovereignty to empirical sovereignty, *Global Governance* 1 (1995): 79–97.

50. T. Homer-Dixon, On the threshold: Environmental changes as causes of acute conflict, *International Security* 12 (1991): 76–116; and Norman Myers, *Ultimate Security: The Environmental Basis of Politics Stability* (New York: Norton, 1993).

51. See, for example, Michel Frédérick, A realist's conceptual definition of environmental security, in Daniel H. Deudney and Richard A. Matthew, eds., *Contested Grounds: Security and Conflict in the New Environmental* Politics (Albany: State University of New York Press, 1999), 91–108.

52. The most significant critiques include Daniel Deudney, The case against linking environmental degradation to national security, *Millennium* 19 (1990): 461–76; Simon Dalby, The politics of environmental security, in J. Kakonen, ed., *Green Security or Militarised Environment* (Dartmouth: Aldershot, 1994); and Jon Barnett, *The Meaning of Environmental Security: Ecological Politics and Policy in the New Security Age* (London: Zed Books, 2001).

53. Dalby, The politics of environmental security, 26.

54. Building on Deudney's critique, Mark Imber has argued that very few ecological problems constitute a genuine *crisis* in the sense of involving a high level of threat, a short period of warning *and* the need for rapid response. Mark Imber, *Environment, Security and UN Reform* (New York: St Martin's Press, 1994), 19.

55. Deudney, "The Case against Linking Environmental Degradation to National Security."

56. Imber, *Environment, Security and UN Reform*, 87.

57. Mathias Finger, Global environmental security and the military, in J. Kakonen, ed., *Green Security or Militarised Environment* (Dartmouth: Aldershot, 1994).

58. Mark Imber, *Environment, Security and UN Reform* (New York: St Martin's Press, 1994), 19.

59. Schrijver, *Sovereignty Over Natural Resources*, 138, n. 54.

60. Steven Bernstein, Liberal environmentalism and global environmental governance, *Global Environmental Politics* 2 (2002): 1–16 at 2. See also Steven Bernstein, *The Compromise of Liberal Environmentalism* (New York: Columbia University Press, 2001).

61. Ken Conca, Rethinking the ecology-sovereignty debate, *Millennium* 23 (1994): 701–11.

62. Robyn Eckersley, Markets, the state and the environment: Introduction and overview, in Robyn Eckersley, ed., *Markets, the State and the Environment:*

Towards Integration (Melbourne: Macmillan, 1995), 7–45. Bernstein traces much of the intellectual defense of these arguments, particularly the polluter pays principle, to the work of the OECD; see Bernstein, Liberal environmentalism and global environmental governance, 8–9.

63. The important question of *how* this has happened is beyond the scope of this chapter, but would entail understanding the actors, arguments and contexts in which new norms emerged, including why certain challenging norms succeeded while others failed. See Jeffrey W. Legro, Which norms matter? Revisiting the "failure of internationalism," *International Organization* 51 (1997): 31–63.

64. World Commission on Environment and Development, *Our Common Future* (Oxford: Oxford University Press, 1987), 348.

65. Neil A. Popovic, In pursuit of environmental human rights: Commentary on the draft declaration of Principles on Human Rights and the Environment, *Columbia Human Rights Law Review* 27 (1995–96): 487–603. See also Hayward, Constitutional environmental rights, 560.

66. Litfin, *The Greening of Sovereignty*, 4.

67. Adriana Fabra, Indigenous peoples, environmental degradation and human rights: A case study, in Alan Boyle and Michael R. Anderson, eds., *Human Rights Approaches to Environmental Protection* (Oxford: Clarendon Press, 1996), 245–64.

68. Transcript of the opening address by Attorney-General Lionel Murphy in the Nuclear Tests Case (Second Phase). *http://lionelmurphy.anu.edu.au/NUCLEAR%20TESTS20%CASE.html* (accessed July 9, 2003). I am grateful to Jenny Hocking for bringing my attention to this address.

69. See Henry Shue, Eroding sovereignty: The advance of principle, in Robert McKim and Jeff McMahan, eds., *The Morality of Nationalism* (New York: Oxford University Press, 1997) for an attempt to reconstruct sovereignty. More typically international political theorists prefer to move beyond traditional sovereignty to an alternative federal system where sovereignty is divided. See, for example, Thomas Pogge, Cosmopolitanism and sovereignty, *Ethics* 103 (1992): 48–75; David Held, *Democracy and the Global Order* (Cambridge: Polity Press, 1995); and Peter Penz, Environmental victims and state sovereignty: A normative analysis, *Social Justice* 23 (1996): 41–61.

70. Shue, Eroding sovereignty.

71. Shue, Eroding sovereignty, 353–54.

72. See Freya Mathews, *The Ecological Self* (London: Routledge, 1991).

73. Penz, Environmental victims and state sovereignty: A normative analysis, 55.

74. Penz, Environmental victims and state sovereignty, 57.

75. Ken Conca, Environmental change and the deep structure of world politics, in Ronnie D. Lipschutz and Ken Conca, eds., *The State and Social Power in*

Global Environmental Politics (New York: Columbia University Press, 1993). See also Conca, Rethinking the ecology-sovereignty debate.

76. Conca, Environmental change and the deep structure of world politics, 311.

77. Penz, Environmental victims and state sovereignty, 56.

78. Yet as Konrad von Moltke has warned in the context of the emerging debate about a possible world environment organization, in view of the complex and vast dimensions of environmental change, "an agency with full responsibility for the environment would wield extraordinary power over all other agencies." Konrad von Moltke, The organization of the impossible, *Global Environmental Politics* 1 (2001): 23–28 at 24.

79. Bull, The state's positive role in world affairs, 112.

Conclusion

1. For a recent defense of the idea that the development of free and critical communication enhances the learning capacity of the social order, see Robert J. Brulle, *Agency, Democracy and Nature: The U.S. Environment Movement from a Critical Theory Perspective* (Cambridge: MIT Press, 2000), especially 272.

2. Hedley Bull, *Justice in International Relations: The Hagey Lectures* (Waterloo, Ontario: University of Waterloo, 1984), 14.

3. Lennart J. Lundqvist, A green fist in a velvet glove: The ecological state and sustainable development, *Environmental Values* 10 (2001): 455–72 at 469.

4. On ecological capacity, see Martin Jänicke and Helmut Weidner, eds., *National Environmental Policies: A Comparative Study of Capacity-Building,* in collaboration with H. Jorgens (Berlin: Springer-Verlag, 1997); and Lundqvist, A green fist in a velvet glove.

5. According to Mikael Skou Andersen and Duncan Liefferink, "It was the spin-off from the simultaneous development of domestic environmental policy in the member states, catalysed by the Stockholm conference, that paved the way for the decision made at the Paris Summit in October 1972. The establishment of EU environmental policy is thus illustrative of the fact that the impact of pioneers was indeed both horizontal and vertical; by their examples, the pioneers served as catalysts for the development of both domestic and international policies." See Mikael Skou Andersen and Duncan Lieferink, Introduction: The impact of the pioneers on EU Environmental Policy, in Mikael Skou Andersen and Duncan Lieferink, eds., *European Environmental Policy: The Pioneers* (Manchester: Manchester University Press, 1997), 5.

6. Jonathan Golub, Introduction and overview, in Jonathan Golub, ed., *Global Competition and EU Environmental Policy* (London: Routledge, 1998).

7. Andrew Hurrell, A crisis of ecological viability? Global environmental change and the nation state, *Political Studies* 42 (1994): 146–65 at 155.

8. Joseph S. Nye, *The Paradox of American Power: Why the World's Only Superpower Can't Go It Alone* (New York: Oxford University Press, 2002).

9. Robert C. Paehlke, *Democracy's Dilemma: Environment, Social Equity, and the Global Economy* (Cambridge: MIT Press, 2003), 272.

10. Lennart J. Lundqvist, Capacity-building or social construction? Explaining Sweden's shift towards ecological modernisation. *GeoForum* 31 (2000): 21–32, 11. See also Lundqvist, Implementation from above: The ecology of power in Sweden's environmental governance, *Governance: An International Journal of Policy and Administration* 14 (2001): 319–37.

11. James Meadowcroft, From welfare state to ecostate, in John Barry and Robyn Eckersley, eds., *The State and the Global Ecological Crisis* (Cambridge: MIT Press, 2004).

Bibliography

Adler, Emmanual. 1997. Seizing the middle ground: Constructivism in world politics. *European Journal of International Relations* 3(3): 319–63.

Adorno, Theodor, and Max Horkheimer. 1972. *The Dialectic of Enlightenment.* New York: Herder.

Andersen, Mikael Skou, and Duncan Liefferink, eds. 1997. Introduction: The impact of the pioneers on EU environmental policy. In *European Environmental Policy: The Pioneers.* Manchester: Manchester University Press.

Anderson, Michael R. 1996. Human rights approaches to environmental protection: An overview. In Alan E. Boyle, and Michael R. Anderson, eds., *Human Rights Approaches to Environmental Protection.* Oxford: Clarendon Press, 1–23.

Anderson, Terry L., and Donald R. Leal. 1991. *Free Market Environmentalism.* San Francisco: Pacific Research Institute for Public Policy.

Archibugi, Daniele, and David Held. 1995. *Cosmopolitan Democracy.* Cambridge: Polity Press.

Arendt, Hannah. 1961. The crisis in culture. In *Between Past and Future: Six Exercises in Political Thought.* New York: Meridan.

Ashley, Richard K. 1984. The poverty of neorealism. *International Organization* 38(2): 225–86.

Axelrod, Robert J. 1984. *The Evolution of Cooperation.* New York: Basic Books.

Bankowski, Zenon, and Emilios Christodoulidis. 1999. Citizenship bound and citizenship unbound. In Kimberly Hutchings, and Roland Dannreuther, eds., *Cosmopolitan Citizenship.* Basingstoke, England: Macmillan, 83–104.

Barkun, J. Samuel, and Bruce Cronin. 1994. The state and the nation: Changing norms and the rules of sovereignty in international relations. *International Organization* 48(1): 107–30.

Barnett, Jon. 2001. *The Meaning of Environmental Security: Ecological Politics and Policy in the New Security Age.* London: Zed Books.

Barnett, Michael. 1995. The new United Nations politics of peace: From juridical sovereignty to empirical sovereignty. *Global Governance* 1(1): 79–97.

Barrow, Clyde W. 1993. *Critical Theories of the State: Marxist, Neo-Marxist, Post-Marxist.* Madison, Wisconsin: University of Wisconsin Press.

Barry, Brian. 1999. Sustainability and Intergenerational Justice. In Andrew Dobson, ed., *Fairness and Futurity: Essays on Environmental Sustainability and Social Justice.* Oxford: Oxford University Press, 93–117.

Barry, John. 1996. Sustainability, political judgment and citizenship: Connecting green politics and democracy. In Brian Doherty, and Marius de Geus, eds., *Democracy and Green Political Thought.* London: Routledge, 115–31.

Barry, John. 1999. *Rethinking Green Politics.* London: Sage Publications.

Barry, John. 2001. Greening liberal democracy: Practice, theory and political economy. In John Barry and Marcel Wissenburg, eds., *Sustaining Liberal Democracy.* Basingstoke, England: Palgrave, 59–80.

Barry, John, and Marcel Wissenburg, eds. 2001. *Sustaining Liberal Democracy.* Basingstoke, England: Palgrave.

Barton, Charmian. 1998. The status of the precautionary principle in Australia: Its emergence in legislation and as a common law doctrine. *Harvard Environmental Law Review* 22: 509–58.

Beck, Ulrich. 1992. *The Risk Society: Towards a New Modernity*, trans. Mark Ritter. London: Sage Publications.

Beck, Ulrich. 1995a. *Ecological Enlightenment: Essays on the Politics of the Risk Society*, trans. Mark A. Ritter. Atlantic Highlands, NJ: Humanities Press.

Beck, Ulrich. 1995b. *Ecological Politics in an Age of Risk*, trans. Amos Weisz. Cambridge: Polity Press.

Beck, Ulrich. 1998. Politics of risk society. In Jane Franklin, ed., *The Politics of the Risk Society.* Cambridge: Polity Press, 9–22.

Beck, Ulrich. 2000. *What Is Globalization?* trans. Patrick Camiller. Cambridge: Polity Press.

Beetham, David. 1991. *The Legitimation of Power.* Atlantic Highlands, NJ: Humanities Press.

Beitz, Charles. 1999. *Political Theory and International Relations*, 2nd ed. Princeton: Princeton University Press.

Bellamy, Richard, and R. J. Barry Jones. 2000. Globalisation and democracy—An afterword. In Barry Holden, ed., *Global Democracy: Key Debates.* London: Routledge, 202–16.

Bernstein, Steven. 2001. *The Compromise of Liberal Environmentalism.* New York: Columbia University Press.

Bernstein, Steven. 2002. Liberal environmentalism and global environmental governance. *Global Environmental Politics* 2(3): 1–16.

Birnie, Patricia. 1995–96. Environment protection and development. *Melbourne University Law Review* 20: 66–100.

Blühdorn, Ingolfur. 2000. Ecological modernization and post-ecological politics. In Gert Spaargaren, Arthur P. J. Mol, and Frederick H. Buttel, eds., *Environment and Global Modernity*, London: Sage, 209–28.

Bohman, James. 1999. The coming of age of deliberative democracy. *Journal of Political Philosophy* 6: 400–25.

Bookchin, Murray. 1982. *The Ecology of Freedom: The Emergence and Dissolution of Hierarchy.* Palo Alto, CA: Cheshire Books.

Boyle, Alan, and Michael R. Anderson, eds. 1996. *Human Rights Approaches to Environmental Protection.* Oxford: Clarendon Press.

Brown, Peter G. 2000. *Ethics, Economics and International Relations: Transparent Sovereignty in the Commonwealth of Life.* Edinburgh: Edinburgh University Press.

Brulle, Robert J. 2000. *Agency, Democracy and Nature: The U.S. Environmental Movement from a Critical Theory Perspective.* Cambridge: MIT Press.

Bull, Hedley. 1979. The state's positive role in world affairs. *Daedalus* 108: 111–23.

Bull, Hedley. 1984. *Justice in International Relations: The Hagey Lectures.* Waterloo, Ontario: University of Waterloo.

Buttel, Frederick H. 1998. Some observations on states, world orders, and the politics of sustainability. *Organisation and Environment* 11(3): 261–86.

Cameron, James, and Ruth Mackenzie. 1996. Access to environmental justice and procedural rights in international institutions. In Alan E. Boyle and Michael R. Anderson, eds., *Human Rights Approaches to Environmental Protection*, Oxford: Clarendon Press, 129–52.

Carr, E. H. 1946. *The Twenty Years' Crisis: An Introduction to the Study of International Relations.* London: Macmillan.

Carter, Alan. 1993. Towards a green political theory. In Andrew Dobson, and Paul Lucardie, eds., *The Politics of Nature: Explorations in Green Political Theory*. London: Routledge, 39–62.

Carter, Alan. 1999. *A Radical Green Political Theory.* London: Routledge.

Cerny, Philip. 1990. *The Changing Architecture of the State: Structure, Agency and the Future of the State.* London: Sage.

Cerny, Philip. 1994. The infrastructure of the infrastructure? 'Toward embedded financial orthodoxy' in the international political economy. In Ronen P. Palan and Barry Gills, eds., *Transcending the State-Global Divide: A Neo-structuralist Agenda in International Relations.* Boulder, CO: Lynne Reinner, 223–49.

Cerny, Philip. 1997. Paradoxes of the competition state: The dynamics of political globalisation. *Government and Opposition* 32(2): 251–74.

Christoff, Peter. 1996. Ecological modernisation, ecological modernities. *Environmental Politics* 5(3): 476–500.

Cohen, Jean L., and Andrew Arato. 1992. *Civil Society and Political Theory.* Cambridge: MIT Press.

Cohen, Joshua. 1997. Procedure and substance in deliberative democracy. In James Bohman, and William Rehg, eds., *Deliberative Democracy: Essays on Reason and Politics*. Cambridge: MIT Press, 426–27.

Commission on Global Governance. 1995. *Our Global Neighbourhood: The Report of the Commission on Global Governance.* Oxford: Oxford University Press.

Conca, Ken. 1993. Environmental Change and the Deep Structure of World Politics. In Ronnie D. Lipschutz, and Ken Conca, eds., *The State and Social Power in Global Environmental Politics.* New York: Columbia University Press, 306–26.

Conca, Ken. 1994. Rethinking the ecology-sovereignty debate. *Millennium* 23(3): 701–11.

Conca, Ken. 2000a. Beyond the statist frame: Environmental politics in a global economy. In Fred P. Gale, and R. Michael M'Gonigle, eds., *Nature, Production, Power: Towards an Ecological Political Economy*. Cheltenham, England: Edward Elgar, 141–55.

Conca, Ken. 2000b. The WTO and the undermining of global environmental governance. *Review of International Political Economy* 7(3): 484–94.

Cox, Robert W. 1981. Social forces, states and world orders: Beyond international relations theory. *Millennium: Journal of International Studies* 10(2): 126–55.

Cox, Robert W. 1995. Critical political economy. In Bjorn Hettne, ed., *International Political Economy: Understanding Global Disorder*. Halifax, Nova Scotia: Fernwood Publishing, 31–45.

Cox, Robert W. 1996. Gramsci, hegemony and international relations. In Robert W. Cox with Timothy J. Sinclair, eds., *Approaches to World Order*. Cambridge: Cambridge University Press.

Cronin, Bruce. 2001. The paradox of hegemony: America's ambiguous relationship with the United Nations. *European Journal of International Relations* 7(1): 103–30.

Dalby, Simon. 1994. The politics of environmental security. In Jyrki Käkönen, ed., *Green Security or Militarised Environment*. Dartmouth: Aldershot, 25–53.

Daly, Herman E., and John B. Cobb Jr. 1989. *For the Common Good: Redirecting the Economy toward Community, the Environment and a Sustainable Future*. Boston: Beacon Press, 145–46.

Delanty, Gerard. 2000. *Citizenship in a Global Age: Society, Culture, Politics.* Houston, JX: Open University Press.

de-Shalit, Avner. 1997. Is liberalism environment-friendly? In Roger S. Gottlieb, ed., *The Ecological Community: Environmental Challenges for Philosophy, Politics and Morality*. London: Routledge, 82–103.

Deudney, Daniel. 1990. The case against linking environmental degradation to national security. *Millennium* 19(3): 461–76.

Deudney, Daniel. 1998. Global village sovereignty: Intergenerational sovereign publics, federal republican earth constitutions, and planetary identities. In Karen T. Litfin, ed., *The Greening of Sovereignty in World Politics*. Cambridge: MIT Press, 299–325.

Deutch, Karl. 1970. *Political Community at the International Level.* New York: Archon Books.

Devetak, Richard, and Richard Higgot. 1999. Justice unbound? Globalisation, states and the transformation of the social bond. *International Affairs* 75(3): 483–98.

Devetak, Richard. 2001. Critical theory. In Scott Burchill, Richard Devetak, Andrew Linklater, Matthew Paterson, Christian Reus-Smit, and Jacqui True, eds., *Theories of International Relations*, 2nd ed. Basingstoke, England: Palgrave, 155–80.

Dobson, Andrew. 1996. Representative democracy and the environment. In *Democracy and the Environment: Problems and Prospects*, edited by William M. Lafferty, and James Meadowcroft. Cheltenham, England: Edward Elgar, 124–39.

Dobson, Andrew. 1998. *Justice and the Environment: Conceptions of Environmental Sustainability and Dimensions of Social Justice.* Oxford: Oxford University Press.

Dobson, Andrew. 2000. *Green Political Thought*, 3rd ed. London: Routledge.

Doherty, Brian, and Marius de Geus, eds. 1996. *Democracy and Green Political Thought: Sustainability, Rights and Citizenship.* London: Routledge.

Dryzek, John S. 1987. *Rational Ecology: Environment and Political Economy.* Oxford: Basil Blackwell.

Dryzek, John S. 1990a. Green reason: Communicative ethics for the biosphere. *Environmental Ethics* 12: 195–210.

Dryzek, John S. 1990b. *Discursive Democracy: Politics, Policy and Political Science.* Cambridge: Cambridge University Press.

Dryzek, John S. 1992. Ecology and discursive democracy: Beyond liberal capitalism and the administrative state. *Capitalism, Nature, Socialism* 3(2): 18–42.

Dryzek, John S. 1997. *The Politics of the Earth: Environmental Discourses.* Oxford: Oxford University Press.

Dryzek, John S. 1999. Transnational democracy. *Journal of Political Philosophy* 7: 30–51.

Dryzek, John S. 2000a. Discursive democracy vs. liberal constitutionalism. In Michael Saward, ed., *Democratic Innovation: Deliberation, Representation and Association.* London: Routledge, 78–89.

Dryzek, John S. 2000b. *Deliberative Democracy and Beyond: Liberals, Critics, Contestations.* Oxford: Oxford University Press.

Dupuy, Pierre. 1997. Due diligence in the international law of liability. In *OECD, Legal Aspects of Transfrontier Pollution*. Paris: OECD, 369–79.

Eckersley, Robyn. 1990. Habermas and green political thought: Two roads diverging. *Theory and Society* 19: 739–76.

Eckersley, Robyn. 1996. Liberal democracy and the rights of nature: The struggle for inclusion. In Freya Matthews, ed., *Ecology and Democracy*. London: Frank Cass, 169–98.

Eckersley, Robyn. 1998. Environmental rights and democracy. In Roger Keil, David Bell, Peter Penz, and Leesa Fawcett, eds., *Political Ecology: Global and Local*. London: Routledge, 353–76.

Eckersley, Robyn. 2000. Disciplining the market, calling in the state: The politics of economy-environment integration. In Stephen Young, ed., *The Emergence of Ecological Modernisation: Integrating the Environment and the Economy*. London: Routledge, 233–52.

Eckersley, Robyn. 2002. Environmental pragmatism, ecocentrism and deliberative democracy: Between problem-solving and fundamental critique. In Bob Pepperman Taylor, and Ben Minteer, eds., *Democracy and the Claims of Nature*. Lanham, MD: Rowman and Littlefield, 49–69.

Eckersley, Robyn. 2004. Soft law/hard politics and the climate change treaty. In Christian Reus-Smit, ed., *The Politics of International Law*. Cambridge: Cambridge University Press.

Elster, Jon. 1998. *Deliberative Democracy.* Cambridge: Cambridge University Press.

Emy, Hugh, and Paul James. 1996. Debating the state: The real reasons for bringing the state back in." In Paul James, ed., *The State in Question*. Sydney: Allen and Unwin, 8–37.

Fabra, Adriana. 1996. Indigenous peoples, environmental degradation and human rights: A case study. In Alan Boyle, and Michael R. Anderson, eds., *Human Rights Approaches to Environmental Protection*. Oxford: Clarendon Press, 245–64.

Falk, Richard. 1997. The critical realist tradition and the demystification of power: E. H. Carr, Hedley Bull and Robert W. Cox. In Stephen Gill, and James H. Mittelman, eds., *Innovation and Transformation in International Studies*. Cambridge: Cambridge University Press, 39–55.

Falk, Richard. 1971. *This Endangered Planet: Prospects and Proposals for Human Survival.* New York: Vintage.

Finger, Matthias. 1994. Global environmental degradation and the military. In Jyrki Käkönen, ed., *Green Security or Militarized Environment?* Aldershot, England: Dartmouth, 169–91.

Fishkin, James. 1995. *The Voice of the People: Public Opinion and Democracy.* New Haven: Yale University Press.

Franck, Thomas. 1990. *The Power of Legitimacy among Nations*. New York: Oxford University Press.

Frédérick, Michel. 1999. A realist's conceptual definition of environmental security. In Daniel H. Deudney, and Richard A. Matthew, eds., *Contested Grounds: Security and Conflict in the New Environmental Politics*. Albany: State University of New York Press, 91–108.

Fukuyama, Francis. 1992. *The End of History and the Last Man*. London: Penguin Books.

Funke, Odelia. 1994. Environmental dimensions of national security. In Jyrki Käkönen, ed., *Green Security or Militarized Environment?* Aldershot, England: Dartmouth, 55–82.

Gargarella, Roberto. 1998. Full representation, deliberation, and impartiality. In Jon Elster, ed., *Deliberative Democracy*. Cambridge: Cambridge University Press, 260–80.

Garratt, Geoffrey. 2000. Globalization and national autonomy. In Ngaire Woods, ed., *The Political Economy of Globalization*. Basingstoke, England: Macmillan, 107–46.

Giddens, Anthony. 1990. *The Consequences of Modernity*. London: Polity Press.

Gill, Stephen. 1995. Globalisation, market civilisation, and disciplinary neoliberalism. *Millennium: Journal of International Studies* 24(3): 399–423.

Goldgeir, James M., and Michael McFaul. 1992. A tale of two worlds: Core and periphery in the post–cold-war era. *International Organization* 46: 467–91.

Goldsmith, Edward. 1988. *The Great U-Turn: De-industrialising Society*. Hartland, England: Green Books.

Golub, Jonathan. 1998. Introduction and overview. In Jonathan Golub, ed., *Global Competition and EU Environmental Policy*. London: Routledge, 1–33.

Goodin, Robert. 1996. Enfranchising the earth, and its alternatives. *Political Studies* 44: 835–49.

Goodin, Robert. 1990. Property rights and preservationist duties. *Inquiry* 33: 401–32.

Grubb, Michael with Michael Vrolijk, and Duncan Brack. 1999. *The Kyoto Protocol: A Guide and Assessment*. London: Royal Institute of International Affairs and Earthscan.

Gurpreet, Rattan. 2001. Prospects for a contemporary republicanism. *Monist* 84(1): 113–30.

Guttman, Amy, and Dennis Thompson. 1996. *Democracy and Disagreement*. Cambridge: Harvard University Press.

Haacke, Jürgen. 1996. Theory and praxis in international relations: Habermas, self-reflection, rational argument. *Millennium* 25(2): 255–89.

Haas, Peter M., Robert O. Keohane, and Mark A. Levy, eds. 1993. *Institutions for the Earth: Sources of Effective International Environmental Protection*. Cambridge: MIT Press.

Habermas, Jürgen. 1971. *Toward a Rational Society: Student Protest, Science and Politics*, trans. Jeremy J. Shapiro. London: Heinemann.

Habermas, Jürgen. 1973. *Legitimation Crisis*, trans. Thomas McCarthy. London: Heinemann.

Habermas, Jürgen. 1987. *The Theory of Communicative Action. Volume Two: Lifeworld and System: A Critique of Functionalist Reason*. Boston: Beacon Press.

Habermas, Jürgen. 1994. Human rights and popular sovereignty: The liberal and republican versions. *Ratio Juris* 7(1): 1–13.

Habermas, Jürgen. 1991. *The Structural Transformation of the Public Sphere: An Inquiry into a Category of Bourgeois Society*, trans. Thomas Burger with the assistance of Frederick Lawrence. Cambridge: MIT Press.

Habermas, Jürgen. 1993. *Justification and Application: Remarks on Discourse Ethics*, trans. Ciaran Cronin. Cambridge: Polity Press.

Habermas, Jürgen. 1996. Three normative models of democracy. In Seyla Benhabib, ed., *Democracy and Difference: Contesting the Boundaries of the Political*. Princeton: Princeton University Press, 21–30.

Habermas, Jürgen. 1997. *Between Facts and Norms*, trans. William Rehg. Cambridge: Polity Press.

Habermas, Jürgen. 1998. *The Inclusion of the Other: Studies in Political Theory*, ed. Ciaran Cronin and Pablo De Greiff. Cambridge: MIT Press.

Habermas, Jürgen. 2001a. A constitution for Europe? *New Left Review* (second series) (September–October): 5–26.

Habermas, Jürgen. 2001b. *The Postnational Constellation: Political Essays*, trans. Max Pensky. Cambridge: Polity Press.

Hajer, Maarten A. 1995. *The Politics of Environmental Discourse: Ecological Modernization and the Policy Process*. Oxford: Clarendon Press.

Hall, Peter A. 1993. Policy paradigms, social learning and the state. *Comparative Politics* 25(3): 275–96.

Hardin, Garrett. 1968. The tragedy of the commons. *Science* 162: 1243–48.

Hasenclever, Andreas, Peter Mayer, and Volker Rittberger. 1997. *Theories of International Regimes*. Cambridge: Cambridge University Press.

Hay, Colin. 1994. Environmental security and state legitimacy. *Capitalism, Nature, Socialism* 5(1): 83–97.

Hay, Colin. 1996. From crisis to catastrophe? The ecological pathologies of the liberal-democratic state. *Innovations* 9(4): 421–34.

Hayward, Tim. 1994. *Ecological Thought: An Introduction*. Cambridge: Polity Press.

Hayward, Tim. 1998. *Political Theory and Ecological Values*. New York: St Martin's Press.

Hayward, Tim. 2000. Constitutional environmental rights: A case for political analysis. *Political Studies* 48: 558–72.

Hayward, Tim. 2004. Greening the constitutional state: Environmental rights in the European Union. In John Barry, and Robyn Eckersley, eds., *The State and the Global Ecological Crisis*. Cambridge: MIT Press.

Held, David, ed. 1991. Democracy and the global system. In *Political Theory Today*. Cambridge: Polity Press, 197–235.

Held, David. 1995. *Democracy and the Global Order.* Cambridge: Polity Press.

Held, David. 1998. Democracy and globalization. In Daniele Archibugi, David Held, and Martin Kohler, eds., *Reimagining Political Community: Studies in Cosmopolitan Democracy*. Cambridge: Polity Press, 11–27.

Held, David. 1999. The transformation of political community: Rethinking democracy in the context of globalization. In Ian Shapiro and Casiana Hacker-Cordón, eds., *Democracy's Edges*. Cambridge: Cambridge University Press, 84–111.

Held, David, Anthony McGrew, David Goldblatt, and Jonathan Perraton. 1999. *Global Transformations: Politics, Economics and Culture.* Stanford, CA: Stanford University Press.

Held, David. 2000. The changing contours of political community: Rethinking democracy in the context of globalization. In Barry Holden, ed., *Global Democracy: Key Debates*, London: Routledge, 17–31.

Hempel, Lamont. 1996. *Environmental Governance: The Global Challenge.* Washington, DC: Island Press.

Hirst, Paul, and Graham Thompson. 1996. *Globalization in Question: The International Economy and the Possibilities of Governance.* Cambridge: Polity Press.

Hirst, Paul. 1997. *From Statism to Pluralism.* London: UCL Press.

Homer-Dixon, Thomas F. 1991. On the threshold: Environmental changes as causes of acute conflict. *International Security* 16(2): 76–116.

Horkheimer, Max. 1947. *The Eclipse of Reason.* New York: Oxford University Press.

Horkheimer, Max. 1972. *Critical Theory: Selected Essays*, trans. Mathew J. O'Connell et al. New York: Seabury Press.

Horton, John. 1992. *Political Obligation.* Basingstoke, England: Macmillan.

Hossain, Kamal and Subrata Roy Chowdhury, eds. 1984. *Permanent Sovereignty Over Natural Resources in International Law: Principles and Practice.* New York: St Martin's Press.

Hurd, Ian. 1999. Legitimacy and authority in international politics. *International Organisation* 53(2): 379–408.

Hurrell, Andrew, and Benedict Kingsbury, eds. 1992. *The International Politics of the Environment.* Oxford: Clarendon Press.

Hurrell, Andrew. 1994. A crisis of ecological viability? Global environmental change and the nation state. *Political Studies* 42: 146–65.

Ikenberry, G. John, and Kupchan, Charles A. 1990. Socialization and hegemonic power. *International Organization* 44(3): 283–315

Imber, Mark. 1994. *Environment, Security and UN Reform.* New York: St Martin's Press.

International Union for the Conservation of Nature and Natural Resources (IUCN). 1980. *World Conservation Strategy: Living Resource Conservation for Sustainable Development*, prepared by the IUCN, with United Nations Environment Programme (UNEP), and the World Wildlife Fund (WWF). Morges, Switzerland: IUCN.

Jackson, Robert H. 1993. *Quasi-States: Sovereignty, International Relations and the Third World.* Cambridge: Cambridge University Press.

Jacobs, Michael. 1999. *Environmental Modernisation: The New Labour Agenda.* London: Fabian Society.

James W. Nickel, and Eduardo Viola. 1994. Integrating environmentalism and human rights. *Environmental Ethics* 16(3): 265–73.

Jänicke, Martin, and Helmut Weidner, eds. 1997. *National Environmental Policies: A Comparative Study of Capacity-Building*, in collaboration with H. Jorgens. Berlin: Springer-Verlag.

Jänicke, Martin. 1990. *State Failure.* Cambridge: Polity Press.

Jänicke, Martin. 1993. Okologische und politische Modernisierung in entwickelten Industriegesellschaften. In V. von Prittwitz, ed., *Umweltpolitik als Modernisierungsprozess. Politikwissenschaftelicke Umwelt-forschung und—lehre in der Bundesrepublik.* Opladen: Leske und Budrich, 15–30.

Johnson, James. 1998. Arguing for deliberation: Some skeptical considerations. In Jon Elster, ed., *Deliberative Democracy.* Cambridge: Cambridge University Press, 161–84.

Johnston, Ronald John. 1989. *Environmental Problems: Nature, Economy and State.* New York: Belhaven Press.

Keck, Margaret, and Kathryn Sikkink. 1998. *Activists beyond Borders: Transnational Issue Networks in International Politics.* Ithaca: Cornell University Press.

Klipsch, Ronald E. 1974. Aspects of a constitutional right to a habitable environment: Towards an environmental due process. *Indiana Law Journal* 49(2): 203–37.

Klotz, Audie. 1995. *Norms in International Relations: The Struggle against Apartheid.* Ithaca: Cornell University Press.

Krasner, Stephen, ed. 1983. *International Regimes.* Ithaca: Cornell University Press.

Krasner, Stephen. 1999. *Sovereignty: Organized Hypocrisy.* Princeton: Princeton University Press.

Kratochwil, Friedrich V. 1989. *Rules, Norms, and Decisions.* Cambridge: Cambridge University Press.

Ksentini, Fatma Zohra. 1994. *Final Report of the UN Subcommission on Human Rights and the Environment.* UN Doc.E/CN.4/Sub.2/1994/9, 6 July.

Kymlicka, Will. 2001. *Politics in the Vernacular: Nationalism, Multiculturalism, and Citizenship.* Oxford: Oxford University Press.

Laclau, Ernesto, and Chantal Mouffe. 1985. *Hegemony and Socialist Strategy: Towards a Radical Democratic Politics.* London: Verso.

Laferrière, Eric. 1996. Emancipating international relations theory: An ecological perspective. *Millennium* 25(1): 53–75.

Laferrière, Eric, and Peter Stoett. 1999. *International Relations Theory and Ecological Thought: Towards a Synthesis.* London: Routledge.

Lammers, J. G., ed. 1987. *Environmental Protection and Sustainable Development: Legal Principles and Recommendations.* London: Graham and Trotman.

Lash, Scott, Bronislaw Szerszynski, and Brian Wynne, eds. 1996. *Risk, Environment and Modernity: Towards a New Ecology.* London: Sage Publications.

Lau, Martin. 1996. Islam and judicial activism: Public interest litigation and environmental protection in the Islamic Republic of Pakistan. In Alan E. Boyle and Michael R. Anderson, eds., *Human Rights Approaches to Environmental Protection.* Oxford: Clarendon Press, 285–302.

Leeson, Susan. 1979. Philosophic implications of the ecological crisis: The authoritarian challenge to Liberalism. *Polity* 11: 305–306.

Legro, Jeffrey W. 1997. Which norms matter? Revisiting the 'failure of internationalism.' *International Organization* 51(1): 31–63.

Light, Andrew, and Eric Katz, eds. 1996. *Environmental Pragmatism.* London: Routledge.

Lindblom, Charles. 1965. *The Intelligence of Democracy: Decision Making Through Mutual Adjustment.* New York: Free Press.

Linklater, Andrew. 1996a. The achievements of critical theory. In Steve Smith, Ken Booth, and Marysia Zalewski, eds., *International Theory: Positivism and Beyond.* Cambridge: Cambridge University Press, 279–98.

Linklater, Andrew. 1996b. Citizenship and sovereignty in the post-Westphalian state. *European Journal of International Relations* 2(1): 77–103.

Linklater, Andrew. 1998. *The Transformation of Political Community: Ethical Foundations of the Post-Westphalian Era.* Cambridge: Polity Press.

Lipschutz, Ronnie D., and Ken Conca, eds. 1993. *The State and Social Power in Global Environmental Politics.* New York: Columbia University Press.

Lisowski, Michael. 2002. Playing the two-level Game: US President Bush's decision to repudiate the Kyoto Protocol. *Environmental Politics* 11(4): 101–19.

Locke, John. 1988. *Two Treatises of Government*, ed. Peter Laslett. Cambridge: Cambridge University Press.

Lundqvist, Lennart J. 2000. Capacity-building or social construction? Explaining Sweden's shift towards ecological modernisation. *GeoForum* 31: 21–32.

Lundqvist, Lennart J. 2001. Implementation from above: The ecology of power in Sweden's environmental governance. *Governance: An International Journal of Policy and Administration* 14(3): 319–37.

Lundqvist, Lennart J. 2001. A green fist in a velvet glove: The ecological state and sustainable development. *Environmental Values* 10(4): 455–72.

Mackay, Michael. 1994. Environmental rights and the US system of protection: Why the US Environmental Protection Agency is not a rights-based administrative agency. *Environment and Planning* 26: 1761–85.

MacIntyre, Alasdair. 1990. The privatization of good: An inaugural lecture. *Review of Politics* 52(3): 344–61.

Marietta, Don E., Jr. 1995. Reflection and environmental activism. In Don E. Marietta, Jr., and Lester Embree, eds., *Environmental Philosophy and Environmental Activism*, Lanham, MD: Rowman and Littlefield, 79–97.

Martínez-Alier, Juan. 1997. Environmental justice (local and global). *Capitalism, Nature, Socialism* 8(1): 91–107.

Matthews, Freya. 1991. *The Ecological Self.* London: Routledge.

Mathews, Freya, ed. 1996. *Ecology and Democracy.* London: Frank Cass.

McGrew, Anthony. 1999. Globalisation and the reconfiguration of political Power. Paper presented before the Civilising the State conference, Deakin University, Melbourne, 5–6 December.

Meadowcroft, James. 1997. Planning for sustainable development: Insights from the literature of political science. *European Journal of Political Research* 31: 427–54.

Meadowcroft, James. 2004. From welfare state to ecostate, in John Barry and Robyn Eckersley, eds., *The State and the Global Ecological Crisis.* Cambridge: MIT Press.

Meadows, Donella H., Dennis L. Meadows, Jörgen Randers, and William H. Behrens III. 1972. *The Limits to Growth.* New York: Universe Books.

Mellor, Mary. 1997. *Feminism and Ecology.* New York: New York University Press.

Mill, J. S. 1979. *Principles of Political Economy*, ed. Donald Winch. Harmondsworth: Penguin.

Miller, David. 1995. *On Nationality.* Oxford: Oxford University Press.

Miller, David. 1999. Bounded citizenship. In Kimberly Hutchings, and Roland Dannreuther, eds., *Cosmopolitan Citizenship.* Basingstoke, England: Macmillan, 60–80.

Miller, David. 2000. In defence of nationality. In John Hutchinson and Anthony D. Smith, eds., *Nationalism: Critical Concepts in Political Science.* London: Routledge, 1676–94.

Minteer, Ben A., and Bob Pepperman Taylor, eds. 2002. *Democracy and the Claims of Nature: Critical Perspectives for a New Century.* Lanham, MD: Rowman and Littlefield.

Mishe, Patricia M. 1993. Ecological scarcity in an interdependent world. In Richard A. Falk, Robert C. Johansen, and Samuel S. Kim, eds., *The Constitutional Foundations of World Peace.* Albany: State University of New York Press, 101–25.

Mol, Arthur A. 1996. Ecological modernisation and institutional reflexivity: Environmental reform in the late modern age. *Environmental Politics* 5(2): 302–23.

Morgenthau, Hans J. 1993. *Politics among Nations: The Struggle for Power and Peace*, brief ed., rev. by Kenneth W. Thompson. New York: McGraw Hill.

Myers, Norman. 1993. *Ultimate Security: The Environmental Basis of Politics Stability.* New York: Norton.

Najam, Adil. 1995. An environmental negotiation strategy for the South. *International Environmental Affairs* 7(3): 249–87.

Nash, Kate. 2000. *Contemporary Political Sociology: Globalization, Politics, and Power.* Malden, MA: Blackwell.

Newell, Peter. 2000. *Climate for Change: Non-state Actors and the Global Politics of Greenhouse.* Cambridge: Cambridge University Press.

Norton, Bryan. 1989. Intergenerational equity and environmental decisions: A model using Rawls' veil of ignorance. *Ecological Economics* 1: 137–59

Norton, Bryan. 1991. *Toward Unity Among Environmentalists.* Oxford: Oxford University Press.

Norton, Bryan. 1996. Integration or reduction: Two approaches to environmental values. In Andrew Light, and Eric Katz, eds., *Environmental Pragmatism.* London: Routledge, 105–38.

Nye, Joseph S. 2002. *The Paradox of American Power: Why the World's Only Superpower Can't Go It Alone.* New York: Oxford University Press.

O'Connor, James. 1998. *Natural Causes: Essays in Ecological Marxism.* New York: Guilford Press.

O'Connor, James. 1973. *The Fiscal Crisis of the State.* New York: St. Martin's Press.

O'Connor, Martin, ed. 1974. On the misadventures of capitalist nature. In *Is Capitalism Sustainable?* New York: Guilford Press.

Oberthür, Sebastian, and Hermann E. Ott. 1999. *The Kyoto Protocol: International Climate Policy for the 21st Century.* Berlin: Springer-Verlag.

OECD. 1994. *Capacity Development in Environment.* Paris: OECD.

Offe, Claus. 1975. The theory of the capitalist state and the problem of policy formation. In Leon N. Lindberg, Robert R. Alford, Colin Grouch, and Claus

Offe, eds., *Stress and Contradiction in Modern Capitalism*. Lexington, MA: D.C. Heath, 125–44.

Offe, Claus. 1984. *Contradictions of the Welfare State*. MIT Press.

Offe, Claus. 1985. *Disorganized Capitalism*. Cambridge: MIT Press.

Offe, Claus. 1996. *Modernity and the State*. Cambridge: MIT Press.

Offe, Claus, and W. von Ronge. Theses on the state. *New German Critique* 6: 139–47.

Ohmae, Kenichi. 1995. *The End of the Nation State: The Rise of the Regional Economies*. New York: Harper Collines/Free Press.

Ophuls, William. 1973. Leviathan or oblivion? In Herman E. Daly, ed., *Toward a Steady State Economy*. San Fransisco: Freeman, 215–30.

Osborne, David, and Ted Gaebler. *Reinventing Government*. Reading, MA: Addison-Wesley.

Oslen, Mancur. 1965. *The Logic of Collective Action*. Cambridge: Harvard University Press.

Ostrom, Elinor. 1990. *Governing the Commons: The Evolution of Institutions for Collective Action*. Cambridge: Cambridge University Press.

Paehlke, Robert. 1988. Democracy, bureaucracy and environmentalism. *Environmental Ethics* 10: 291–308.

Paehlke, Robert C. 2003. *Democracy's Dilemma: Environment, Social Equity, and the Global Economy*. Cambridge: MIT Press.

Palan, Ronan, and Abbott, Jason. 1996. *State Strategies in the Global Economy*. London: Pinter.

Pallemaerts, Marc. 1993. International Environmental Law from Stockholm to Rio: Back to the Future? In Philippe Sands, ed., *Greening International Law*. London: Earthscan, 1–19.

Parekh, Bhikhu. 1992. The cultural particularity of liberal democracy. *Political Studies* 40 (Special issue): 160–75.

Paterson, Matthew. 1999. Green political strategy and the state. In N. Ben Fairweather, Sue Elworthy, Matt Stroh, and Piers H. G. Stephens, eds., *Environmental Futures*. Basingstoke, England: Macmillan, 73–87.

Payne, Rodger. 2000. Habermas, discourse norms, and the prospects for global deliberation. Paper presented before Annual Meeting, International Studies Association, Los Angeles, 15–18 March.

Penz, Peter. 1996. Environmental victims and state sovereignty: A normative analysis. *Social Justice* 23(4): 41–61.

Perrez, Franz Xavier. 1996. The relationship between "permanent sovereignty" and the obligation not to cause transboundary environmental damage. *Environmental Law* 26: 1187–1212.

Petitt, Philip. 1997. *Republicanism: A Theory of Freedom and Government*. Oxford: Oxford University Press.

Phillips, Anne. 1996. Dealing with difference: A politics of ideas or a politics of presence? In Seyla Benhabib, ed., *Democracy and Difference: Contesting the Boundaries of the Political.* Princeton: Princeton University Press, 139–52.

Philpott, Daniel. 2001. *Revolutions in Sovereignty: How Ideas Shaped Modern International Society.* Princeton: Princeton University Press.

Plumwood, Val. 1991. Nature, self, and gender: Feminism, environmental philosophy, and the critique of rationalism. *Hypatia* 6(1): 3–27.

Plumwood, Val. 1993. *Feminism and the Mastery of Nature.* London: Routledge.

Plumwood, Val. 1996. Has democracy failed ecology? An ecofeminist perspective. *Environmental Politics* 4(4): 134–68.

Plumwood, Val. 2002. *Environmental Culture: The Ecological Crisis of Reason.* London: Routledge.

Pogge, Thomas. 1989. *Realizing Rawls.* Ithaca: Cornell University Press.

Pogge, Thomas. 1992. Cosmopolitanism and sovereignty. *Ethics* 103: 48–75.

Poggi, Gianfranco. 1990. *The State: Its Nature, Development and Prospects.* Stanford: Stanford University Press.

Polkinghorne, Donald. 1988. *Narrative Knowing and the Human Sciences.* Albany: State University of New York Press.

Popovic, Neil A. F. 1995–96. In pursuit of environmental human rights: Commentary on the draft declaration of Principles on Human Rights and the Environment. *Columbia Human Rights Law Review* 27(3): 487–603.

Porter, Gareth, Janet Welsh Brown, and Pamela S. Chasek. 2000. *Global Environmental Politics*, 2nd ed. Boulder, CO: Westview Press.

Price, Richard, and Christian Reus-Smit. 1998. Dangerous liaisons: Critical international theory and constructivism. *European Journal of International Relations* 4(3): 259–94.

Price, Richard. 1997. Moral norms in world politics. *Pacific Review* 9(1): 45–72.

Pulido, L. 1994. Restructuring and the contraction and expansion of environmental rights in the United States. *Environment and Planning* A26: 915–36.

Rees, William, and Mathis Wackernagel. 1994. Ecological footprints and appropriated carrying capacity. In AnnMari Jansson, Monica Hammer, Carl Folke, and Robert Constanza, eds., *Investing in Natural Capital: The Ecological Economics Approach to Sustainability.* Washington, DC: Island Press, 362–90.

Rehg, William. 1999. Intractable conflicts and moral objectivity: A dialogical, problem-based approach. *Inquiry* 42: 229–58.

Reus-Smit, Christian. 1996. The normative structure of international society. In Fen Osler Hampson, and Judith Reppy, eds., *Earthly Goods: Environmental Change and Social Justice.* Ithaca: Cornell University Press, 96–121.

Reus-Smit, Christian. 1999. *The Moral Purpose of the State.* Princeton: Princeton University Press.

Reus-Smit, Christian. 2001. Human rights and the social construction of sovereignty. *Review of International Studies* 27: 519–38.

Risse, Thomas. 2000. "Let's argue!": Communicative action in world politics. *International Organisation* 54(1): 1–39.

Rosewarne, Stuart. 1997. Marxism, the second contradiction, and socialist ecology. *Capitalism, Nature, Socialism* 8(2): 99–120.

Ruggie, John Gerard. 1993. Territoriality and beyond: Problematizing modernity in international relations. *International Organisation* 47(1): 144–73.

Ruggie, John Gerard. 1998. *Constructing the World Polity: Essays on International Institutionalization.* London: Routledge.

Rutherford, Paul. 1999. Ecological modernisation and environmental risk. In Éric Darier, ed., *Discourses of the Environment.* Oxford: Blackwell, 95–118.

Sachs, Wolfgang, ed. 1993. *Global Ecology.* London: Zed Books.

Sagoff, Mark. 1988. *The Economy of the Earth.* Cambridge: Cambridge University Press.

Said, Edward. 1989. Representing the colonized: Anthropology's interlocutors. *Critical Inquiry* 15: 205–225.

Samhat, Nayef. 1997. International regimes as political community. *Millennium* 26(2): 349–78.

Sandel, Michael. 1982. *Liberalism and the Limits of Justice.* Cambridge: Cambridge University Press.

Sanders, Lynne. 1997. Against deliberation. *Political Theory* 25: 347–76.

Sassen, Saskia. 1996. *Losing Control? Sovereignty in an Age of Globalization.* New York: Columbia University Press.

Saurin, Julian. 1996. International relations, social ecology and the globalisation of environmental change. In John Vogler and Mark Imber, eds., *The Environment and International Relations.* London: Routledge, 77–98.

Saward, Michael. 1998. Green state/democratic state. *Contemporary Politics* 4(4): 345–56.

Saward. Michael. 2000. A critique of Held. In Barry Holden, ed., *Global Democracy: Key Debates.* London: Routledge, 37–38.

Schnaiberg, Alan. 1980. *The Environment.* New York: Oxford University Press.

Schattschneider, Eric E. 1960. *The Semisovereign People: A Realist's View of Democracy in America.* New York: Holt, Rinehart and Winston.

Scholte, Jan Aart. 2000. Global civil society. In Ngaire Woods, ed., *The Political Economy of Globalization.* Basingstoke, England: Macmillan, 173–201.

Schrijver, Nico. 1997. *Sovereignty over Natural Resources: Balancing Rights and Duties.* Cambridge: Cambridge University Press.

Shaw, Martin. 2000. *Theory of the Global State: Globality as an Unfinished Revolution.* Cambridge: Cambridge University Press.

Shelton, Dinah. 1991. Human rights, environmental rights and the right to environment. *Stanford Journal of International Law* 28(39): 103–38.

Shiva, Vandana. 1988. *Staying Alive: Women, Ecology and Development.* London: Zed Books.

Shue, Henry. 1995. Ethics, the environment and the changing international order. *International Affairs* 71(3): 453–61.

Shue, Henry. 1997. Eroding sovereignty: The advance of principle. In Robert McKim and Jeff McMahan, eds., *The Morality of Nationalism.* New York: Oxford University Press.

Skocpol, Theda. 1979. *States and Social Revolutions.* Cambridge: Cambridge University Press.

Skocpol, Theda. 1985. Bringing the state back in: Strategies of analysis in current research. In Peter Evans, Dietrich Reuschemeyer, and Theda Skocpol, eds., *Bringing the State Back In.* Cambridge: Cambridge University Press, 3–37.

Smith, Steve. 1993. The environment on the periphery of international relations. *Environmental Politics* 2(4): 28–45.

Soper, Kate. 1995. *What Is Nature?* Oxford: Blackwell.

Sorensen, Georg. 1999. Sovereignty: Change and continuity in a fundamental institution. *Political Studies* 47: 590–604

Sprinz, Detlef, and Tapani Vaahtoranta. The interest-based explanation of international environmental policy. *International Organization* 48(1): 77–105.

Strange, Susan. 1995. The defective state. *Daedalus* 124 (Spring): 55–74.

Sunstein, Cass R. 1997. Deliberation, democracy and disagreement. In Ron Bontekoe, and Marietta Stepaniants, eds., *Justice and Democracy: Cross-Cultural Perspectives.* Honolulu: University of Hawaii Press, 93–117.

Tarlock, A. Dan. 1988. Earth and other ethics: The institutional issues. *Tennessee Law Review* 56: 43–76.

Taylor, Charles. 1985. Atomism. In *Philosophy and the Human Sciences: Philosophical Papers 2.* Cambridge: Cambridge University Press, 87–210.

Taylor, Charles. 1992. *Multiculturalism and the Politics of Recognition.* Princeton: Princeton University Press.

Taylor, Michael. 1987. *The Possibility of Cooperation.* Cambridge: Cambridge University Press.

Taylor, Prue. 1998. *An Ecological Approach to International Law: Responding to Challenges of Climate Change.* London: Routledge.

Thatcher, Margaret. 1992. Speech to Scottish Conservative Party Conference, 14 May 1982. Reproduced in *The Oxford Dictionary of Quotations*, ed. A. Partington. 4th ed.

Thompson, Dennis. 1999. Democratic theory and global society. *Journal of Political Philosophy* 7: 111–25.

Tilly, Charles. 1985. War making and state making as organized crime. In Peter Evans, Dietrich Reuschemeyer, and Theda Skocpol, eds., *Bringing the State Back In*. Cambridge: Cambridge University Press, 169–91.

Torgerson, Douglas. 1997. Policy professionalism and the voices of dissent: The case of environmentalism. *Polity* 29(3): 345–74.

Torgerson, Douglas. 1999. *The Promise of Green Politics: Environmentalism and the Public Sphere*. Durham: Duke University Press.

Torgerson, Douglas. 2000. Farewell to the green movement? Political action and the green public sphere. *Environmental Politics* 9(4): 1–19.

Uhr, John. 1993. Instituting republicanism: Parliamentary vices, republican virtues. *Australian Journal of Political Science* (Special issue) 28: 27–39.

Varner, Gary E. 1994. Environmental law and the eclipse of land as private property. In Frederick Ferré and Peter Hartel, eds., *Ethics and Environmental Policy: Theory Meets Practice*. Athens: University of Georgia Press, 142–60.

Vig, Norman. 1997. Toward common learning: Trends in US and EU environmental policy. Lecture for Summer Symposium on Innovation of Environmental Policy, Bologna, 22 July.

Vincent, Andrew. 1998. Liberalism and the environment. *Environmental Values* 7(4): 443–59.

Vincent, Andrew. 1987. *Theories of the State*. Oxford: Blackwell.

Vogel, Steven. 1995. *Against Nature: The Concept of Nature in Critical Theory*. Albany: State University of New York Press.

Vogel, Steven. 1997. Habermas and the ethics of nature. In Roger Gottlieb, ed., *The Ecological Community*. New York: Routledge, 175–92.

Vogler, John. 1996. Introduction. The environment in international relations: Legacies and contentions. In John Vogler and Mark Imber, eds., *The Environment and International Relations*. London: Routledge, 1–21.

von Moltke, Konrad. 2001. The organisation of the impossible. *Global Environmental Politics* 1(1): 23–28.

von Weizsacker, Ernst, Amory B. Lovins, and L. Hunter Lovins. 1997. *Factor 4: Doubling Wealth—Halving Resource Use*. Sydney: Allen and Unwin.

Walker, Ken. 1989. The state in environmental management: The ecological dimension. *Political Studies* 37: 25–39.

Walzer, Michael. 1994. *Thick and Thin: Moral Argument at Home and Abroad*. Notre Dame: University of Notre Dame Press.

Weale, Albert. 1992. *The New Politics of Pollution*. Manchester: Manchester University Press.

Weber, Max. 1948. Politics as vocation. In *From Max Weber: Essays in Sociology*, trans. and ed. H. H. Gerth and C. Wright Mills. London: Routledge and Kegan Paul.

Weiss, Edith Brown. 1993. International environmental law: Contemporary issues and the emergence of a new world order. *Georgetown Law Journal* 81 (March): 675–710.

Weiss, Linda. 1998. *The Myth of the Powerless State: Governing the Economy in a Global Era.* Cambridge: Polity Press.

Wendt, Alexander. 1992. Anarchy is what states make of it: The social construction of power politics. *International Organization* 46(2): 391–426.

Wendt, Alexander. 1999a. *Social Theory of International Politics.* Cambridge: Cambridge University Press.

Wendt, Alexander. 1999b. A comment on Held's cosmopolitanism. In Ian Shapiro and Casiana Hacker-Cordón, eds., *Democracy's Edges.* Cambridge: Cambridge University Press, 127–33.

Wheeler, David. 2001. Racing to the bottom? Foreign investment and air pollution in developing countries. *Journal of Environment and Development* 10(3): 225–45.

Wissenburg, Marcel. 1998. *Green Liberalism: The Free and the Green Society.* London: UCL Press.

Woods, Ngaire, ed. 2000. The political economy of globalization. In *The Political Economy of Globalization.* Basingstoke, England: Macmillan, 1–19.

World Commission on Environment and Development. 1987. *Our Common Future: The Report of the World Commission on Environment and Development.* Oxford: Oxford University Press.

Wyn Jones, Richard, ed. 2001. Introduction: Locating critical international relations theory. In *Critical Theory and World Politics.* Boulder, CO: Lynne Rienner, 1–19.

Young, Iris Marion. 1996. Communication and the other: Beyond deliberative democracy. In Seyla Benhabib, ed., *Democracy and Difference: Contesting the Boundaries of the Political.* Princeton: Princeton University Press, 120–35.

Young, Iris Marion. 1997. Difference as a resource for democratic communication. In James Bohman and William Rehg, eds., *Deliberative Democracy: Essays on Reason and Politics.* Cambridge: MIT Press, 383–406.

Young, Iris Marion. 2000. *Democracy and Inclusion.* Oxford: Oxford University Press.

Young, Oran R. 1994. *International Governance: Protecting the Environment in a Stateless Society.* Ithaca: Cornell University Press.

Zolo, Danilo. 2000. The lords of peace: From the holy alliance to the new international criminal tribunals. In Barry Holden, ed., *Global Democracy: Key Debates.* London: Routledge, 73–86.

Zürn, Michael. 1998. The rise of international environmental politics. *World Politics* 50: 617–49.

Cases

Trail Smelter Case (United States *v.* Canada) 3 R International Arbitration Awards 1905.

Corfu Channel Case (Merits) (United Kingdom *v.* Albania) [1949] ICJ Reports 4.

Lac Lanoux Arbitration (Spain *v.* France) 12 R International Arbitration Awards 281.

Nuclear Tests Case (Australia *v.* France) [1973] *ICJ Reports* 99; (New Zealand *v.* France) [1974] *ICJ Reports* 135.

Transcript of the opening address by Attorney-General Lionel Murphy in the Nuclear Tests Case (Second Phase).

http://lionelmurphy.anu.edu.au/NUCLEAR%20TESTS%CASE.html

Index